Earthly Words

Center for Environmental Arts and Humanities
Department of English/098
University of Nevada, Reno
Reno, NV 89557 USA

Earthly Words

Essays on Contemporary American Nature and Environmental Writers

Edited by John Cooley

Ann Arbor
The University of Michigan Press

Copyright © by the University of Michigan 1994
All rights reserved
Published in the United States of America by
The University of Michigan Press
Manufactured in the United States of America
∞ Printed on acid-free paper

1997 1996 1995 1994   4 3 2 1

A CIP catalogue record for this book is available from the British Library.

Library of Congress Cataloging-in-Publication Data

Earthly words : essays on contemporary American nature and
    environmental writers / edited by John Cooley.
        p.   cm.
    Includes bibliographical references.
    ISBN 0-472-09537-4 (alk. paper) — ISBN 0-472-06537-8 (pbk. :
alk. paper)
    1. American literature—20th century—History and criticism.
2. Natural history—United States—Historiography. 3. Nature in
literature. 4. Environmental protection in literature. I. Cooley,
John R., 1937–   .
PS163.E2 1994
810.9'36'09045—dc20   94-15223
                                                                    CIP

In memory of Frank Harder Ryder 1878–1969
—he led his grandson to the wild country

Preface

This volume brings together a selection of critical essays that focus on contemporary American nature and environmental writers. Nature writing is flourishing in America as never before, including distinguished fiction and some of the finest prose nonfiction being written today. The essays included point to the growing academic attention these writers are receiving in classrooms, in major anthologies (such as the recent *Norton Book of Nature Writing*), and in scholarly journals. In addition to academic stature, contemporary nature writers enjoy the unusual distinction of a growing popularity with general audiences. This broadly based attention reflects more than a healthy popular interest in and desire to better understand natural systems; it also registers a growing concern of thoughtful Americans over the deterioration of environmental conditions and desire to better understand natural systems.

Earthly Words provides an overview of the careers and publications of America's most significant and influential contemporary nature writers. Most of the essays included in this collection give an introduction to each writer's major publications, themes, and achievements. Other essays contribute in-depth interpretations of selected texts, such as Abbey's *Desert Solitaire,* Dillard's *A Pilgrim at Tinker Creek,* and Leopold's *A Sand County Almanac,* texts that have achieved notable popularity and critical standing and that have contributed greatly to public appreciation of nature and ecology. Each author section also includes a brief biographical sketch of the author in question and, at the end of the volume, a primary and secondary bibliography.

In selecting the writers to be represented in this collection, the editor was guided and constrained by the amount of critical attention received by an individual writer, the possibility of commissioning or reprinting essays, and the projected size of the edition. As a result, some accomplished contemporary nature writers, such as Edward Hoagland, Gretel Ehrlich, Rachel Carson, Lewis Thomas, and Farley Mowat, could not be included.

The Introduction briefly traces the literary and intellectual history of nature writing and the pastoral, the major ideas the pastoral expresses, and some of the forms it has taken. Despite the great differences in subject and style among the writers given scrutiny here, they share a common tradition of inherited forms, themes, and conventions, as the Introduction makes clear. They also convey their knowledge of and affection for the natural world as well as direct and implied critiques of human values and activities. They ask: How shall we live? What place shall nonhuman nature occupy in our lives? How can we find more rewarding and harmonious relationships with the ecosystem that supports all life? As if a spokesperson for nature writers, Joseph Wood Krutch expresses the common focus and concern this way: "we are all in this together."

Acknowledgments

This book was conceived in the "Nature and Environmental Writers" sections of the Popular Culture Association–American Culture Association conferences. I extend my appreciation to members of this circle for their many stimulating papers and discussions about American nature writers, especially to Professors Adelia Peters (Bowling Green State University), Walter Herrscher (University of Wisconsin–Green Bay), James McClintock (Michigan State University), and Thomas Bailey (Western Michigan University) for their knowledge and good advice. I must also acknowledge valuable conversations with Wendell Berry, Barry Lopez, and Peter Matthiessen, about nature writing and about their own work. Without the generous support of Western Michigan University in granting me a sabbatical leave to work on this project, it would not have been accomplished. My appreciation extends to the English Department and the Environmental Studies Program for collegial and secretarial support, and for the invaluable assistance of graduate students John Hanley, Daniel Burns, and Sandy Stafford. Other colleagues who have given valuable advice and encouragement include William Rueckert, Bradley Hayden, Robert Haight, and Peter Schmidt, and American Studies colleagues at the University of Nottingham.

Grateful acknowledgment is made to the following publishers and individuals for use of copyrighted material:

Farrar, Straus & Giroux for "A Few Words in Favor of Edward Abbey," from *What Are People For,* copyright © 1990 by Wendell Berry. Published by North Point Press and reprinted by permission of Farrar, Straus & Giroux, Inc.

Ed Folsom for "Gary Snyder's Descent to Turtle Island: Searching for Fossil Love," which originally appeared in *Western American Literature* 15 (1980): 103–21. Reprinted by permission of the author.

Jack Hicks, "Wendell Berry's Husband to the World: A Place on Earth," *American Literature* 51:2 © 1979 Duke University Press. Reprinted with permission of the publisher.

Acknowledgments

Henry Holt and Company for "Mr. Krutch" from *One Life at a Time, Please*, by Edward Abbey. Copyright © 1978, 1983, 1985, 1986, 1988 by Edward Abbey. Reprinted by permission of Henry Holt and Company, Inc.

Gary McIlroy, "*Pilgrim at Tinker Creek* and the Social Legacy of *Walden*," *South Atlantic Quarterly* 85:2, 1986, copyright Duke University Press. Reprinted with permission of the publisher.

Patrick D. Murphy for "Penance and Perception: Spirituality and Land in the Poetry of Gary Snyder and Wendell Berry," which first appeared in *Sagtrieb*, Fall (1986): 61–72. Reprinted by permission of the author.

William H. Rueckert for "Barry Lopez and the Search for a Dignified and Honorable Relationship with Nature," from the *North Dakota Quarterly* 59 (Spring 1991): 279–304. © 1991.

University of Wisconsin Press for "Anatomy of a Classic," by John Tallmadge, which first appeared in *Companion to Sand County Almanac: Interpretive and Critical Essays*, edited by Baird Callicott, © 1987 by University of Wisconsin Press.

Diane Wakoski for "Edward Abbey: Joining the Visionary Inhumanists," from *Resist Much, Obey Little: Some Notes on Edward Abbey*, edited by James Hepworth, and Gregory McNamee (Dream Garden Press, 1985). Reprinted by permission of the author.

Stephen Weiland for "Wendell Berry Resettles America: Fidelity, Education, and Culture," which first appeared in *Iowa Review* 10, no. 1 (1979): 99–104. Copyright © 1979.

Every effort has been made to trace the ownership of all copyrighted material in this book and to obtain permission for its use.

Contents

Preface   vii

Acknowledgments   ix

Introduction: American Nature Writing and the
   Pastoral Tradition   1
John Cooley

EDWARD ABBEY   17

Wendell Berry
   A Few Words in Favor of Edward Abbey   19
Diane Wakoski
   Edward Abbey: Joining the Visionary "Inhumanists"   29

WENDELL BERRY   35

Steven Weiland
   Wendell Berry Resettles America: Fidelity, Education,
   and Culture   37
Jack Hicks
   Wendell Berry's Husband to the World:
   A Place on Earth   51

ANNIE DILLARD   67

James I. McClintock
   "Pray Without Ceasing": Annie Dillard among the
   Nature Writers   69
Gary McIlroy
   Pilgrim at Tinker Creek and the Social Legacy
   of Walden   87

JOSEPH WOOD KRUTCH   103
   Edward Abbey
      "Mr. Krutch"   105

ALDO LEOPOLD   117
   John Tallmadge
      Anatomy of a Classic   119

BARRY LOPEZ   135
   William H. Rueckert
      Barry Lopez and the Search for a Dignified and Honorable Relationship with Nature   137

PETER MATTHIESSEN   165
   John Cooley
      Matthiessen's Voyages on the River Styx: Deathly Waters, Endangered Peoples   167

JOHN MCPHEE   193
   Thomas C. Bailey
      John McPhee: The Making of a Meta-Naturalist   195

GARY SNYDER   215
   Ed Folsom
      Gary Snyder's Descent to Turtle Island: Searching for Fossil Love   217
   Patrick D. Murphy
      Penance or Perception: Spirituality and Land in the Poetry of Gary Snyder and Wendell Berry   237

Afterword: Toward an Ecocriticism   251
   John Cooley

Contributors   255

Bibliographies   257

Introduction: American Nature Writing and the Pastoral Tradition

John Cooley

This book focuses on texts written since 1945 by contemporary American nature and environmental writers. Because a number of the essays included make reference to earlier nature writing, much of it written in the pastoral mode, this introduction defines nature writing in general and its important subsidiary the pastoral and gives a brief summary of European development of the pastoral. It then describes the development of nature writing, including pastoralism, in America, and suggests ways of relating the essays in this book and their subjects to this historical context.

Nature Writing and the European Development of the Pastoral

In the broadest perspective, nature writing refers to all texts that describe or study nonhuman environments, including texts that examine the interactions between such environments and humans. Nature writing embraces all the creative literary forms and the full range of nonfiction prose. This latter category contains a remarkably diverse range of writings from the biological and earth sciences—including taxonomies, field guides, and ecological studies—and from the humanities—including natural histories, environmental journalism, essays, and travel writing. It also contains more personal forms of nature writing, such as autobiographies, diaries, herbals, bestiaries, and country life journals.

The pastoral is, technically speaking, a mode of writing rather than a type, and thus it appears sometimes in unexpected situations. But the pastoral is rarely found in ecological texts, field guides, scientific studies, and environmental journalism. It has to do more directly with the interaction of individual humans and nature, whether in permanent residence or in brief sojourn into wilderness or farmland settings. Regardless of the genre in which they are written, pastorals usually contain a narrative about country or wilderness life. Typically, the pastoral strongly identifies with place, with the local geography and natural history, and it often exhibits an awareness of regional ecology, including such significant features as soils, habitat, seasonality, and climate.

Pastorals frequently present a spokesperson or guide who is knowledgeable of and represents the local landscape. Such a figure, whether connected with farming or herding, with the agricultural or the wilder landscapes, serves by example, and sometimes by invitation, as guide to urban adventurers, and vicariously to urban readers. (Historically, urban dwellers have constituted the primary readership for pastorals and for nature writing in general.) Pastoral figures, guides, companions, and even reluctant providers of direction or assistance give the outsider factual information and, on some occasions, ecological understanding and spiritual insight. Pastoral guides often help urban outsiders find their way and may help them sharpen their senses and reflexes to become better-informed "readers" of nature. Readers of pastorals may emerge with a heightened understanding of the delicate balances that must be struck between country life and nature's needs, between human demands for natural resources and damage caused by human ignorance of or indifference to ecosystem needs and limits. The pastoral mode is wide-ranging and versatile: Shakespeare's *The Tempest* fits the guidelines, as do many poems by Wordsworth and Robert Frost. Henry Thoreau's *Walden: Or Life in the Woods* is generally regarded as the consummate American pastoral.

The origins of the pastoral as a literary convention are usually attributed to the Greek poet Theocritus (third century B.C.), who depicted in his *Idylls* the agreeable and happy life of Sicilian shepherds. As a broader cultural form, pastoralism, which is in essence an attitude of positive regard for herdsmen or country people, their habits of life and their outlook, has been traced to Mesopotamia as early as 3100 B.C. In the ancient Near East, the pastoral figure moved back and forth between settled society, agricultural settings, and unmodified wild lands and evolved into a mediator between these realms. Even though the pastoral figure's first concern had to be the protection of herds from natural dangers, the figure also came to epitomize, as Leo Marx expresses it, "... the virtues of a simple unworldly life, disengaged from civilization and lived, as we now might say, 'close to nature.'"[1]

Despite numerous forms and versions of pastoralism, the central elements have been relatively constant: the rewarding, although usually temporary, removal from city to country, and the reliance on the herdsman or pastoral spokesperson to express nature's wisdom and to mediate between "the constraints of society and the constraints of nature."[2] For some literary historians, the Georgic tradition is significantly different from that of the pastoral, recognizing specifically the culture of settled agriculture, as opposed to that of the more remote and nomadic herdsman. Because this distinction is seldom maintained in American nature writing and its criticism, I extend my general concept of the pastoral to embrace the Georgic as well.

The Roman poet Virgil revived and expanded the pastoral in his *Eclogues* and set the pastoral convention that has been followed by centuries of European poets—an expressed longing on the part of an urbane narrator for a life of peace and simplicity in an agreeable country setting. At the heart of classical pastoralism, as a literary mode, is an idealization of an earlier time and paradisiacal condition that are no longer achievable except in the imagination. Thus, pastorals are frequently tinged with nostalgia for or regret over the loss of an idyllic condition: childhood, a perfect love, an idealized farm, a promised land, the innocence of Eden. Indeed, such prelapsarian fantasy is a recognizable strand in the pastoral. Greek and Roman writers yearned for an age of gold, a perfection that further eluded humanity with each passing age. Christian pastoralists expressed this root idea in a specific yearning for the restoration of the Garden of Eden as the perfection of the pastoral. Consequently, images of deterioration, decay, and decadence play seductive but significant roles even for contemporary nature writers attracted to the pastoral, such as Peter Matthiessen and Wendell Berry. Such writers convey frequent images of decay, exile, and loss of a promised or ideal land.[3] These losses are almost always attributed to human greed, rapacity, or stupidity.

Although the pastoral has tended toward nostalgia for the past and an idealization of country and farming life, it is not so straightforward a concept as it may first appear. As Harold Tolliver has pointed out, the evocation of a pastoral ideal "habitually calls forth its opposite."[4] Tolliver identifies elements of conflict in pastorals from Virgil's *Eclogues* (fear of losing farmland to encroaching Rome) to Saul Bellow's *Henderson the Rain*

King (impact of industrial capitalism on the human psyche). Even in pastorals designed primarily to soothe and entertain, or to indulge in sentimental country retreats, underlying elements of conflict often await the discerning reader—conflicts between the ideal and the real, or perhaps between the static past and the changing present. The best pastorals have challenged and informed readers even while entertaining them with the pleasures and the harsh beauty of nature and country life.

William Empson, in his classic study *Some Versions of Pastoral,* observes that at the heart of all complex pastorals is the act of "putting the complex into the simple."[5] The pastoral attempts to maintain or restore a balance between nature and civilization, he argues, by elevating the shepherd, the farmer, the country person and by subverting the usual heroic convention that elevates the civilized power of the aristocracy, the clergy, political and military leaders. The pastoral writer, in Empson's view, has a strong proletarian bias, attempting to speak for the powerless, those of the land who are not well represented in educated and powerful circles. Traditionally, pastoral writers express farm and country ways and values to a civilized, urbane readership, although they have seldom been fully accepted in either camp. Many pastoral poems have expressed simplistic and sentimental, urbane views about farming and country life. But the pastoral mode also contains numerous complex texts—by poets as diverse as Virgil, Wordsworth, and Wendell Berry—that approach the realities of country life from a country perspective. These tough-minded and complex pastorals express acute awareness of landscape zones—urban, farm, and wilderness—and of such issues as land ownership, economics, and political power. A majority of the writers receiving critical commentary in this edition write in the vein of the complex pastoral.

Recent British pastoral criticism, following William Empson's work, has been particularly interested in issues of class, land ownership, politics, and economics—all the ingredients for a rigorous political criticism. Notable among these critics is Williams for his study *The Country and the City.* Williams observes that the commonly expressed environmental concern for a disappearing rural and natural world is, no matter how valid, a theme as old as the earliest pastorals. As a Marxist critic, Williams is well situated for his extensive critique of the impact of capitalism on both city and country in the industrial societies of the West. He argues that capitalism, as a mode of production, has shaped almost all that we see and know as the history of city and country in Europe and America: "Its abstracted economic drives, its fundamental priorities in social relations, its criteria of growth and of profit and loss, have over several centuries altered our country and created our kinds of city." Looking at the many disastrous consequences to both human society and the natural world of our profit-first, free-enterprise value system, Williams argues that, short of open opposition, a persistent "resistance to capitalism" is the superior course of action.[6]

More recently, Roger Sales has applied the Empson–Williams line of political pastoral criticism to English literature and society during the Romantic period. In *English Literature and History 1778–1830*, he argues that the pastoral has never been a naive and innocent literary convention in England, and that in certain, subtle ways, it justified existing patterns of landownership and supported established religious and political authority. In Sales's view, most English pastorals support the rural status quo. As he expresses it, "the pastoral perspective takes the form of a blurred long-shot rather than a tight close-up."[7]

Sales is part of a revisionist critical position that emerged during the 1980s and that has explored the ideological dimensions of English Romanticism. Sales, Geoffrey Hartman, Harold Bloom, Jerome McGann, and others have argued, in general, that the Romantics, especially Wordsworth, privileged the individual imagination over history, politics, and economics. In so doing, the argument runs, the Romantics ignored social conditions in their quest for the transcendent sublime. (The same criticism has been made of several American nature writers, particularly Annie Dillard.) Sales argues that English pastorals habitually arrive too late to be of polemic usefulness in clarifying and shaping public opinion. The English pastoral, in his view, has never been an effective vehicle for revolution, reform, or environmental protection. By contrast, American pastoral writers are currently addressing some of these often-ignored social issues. Wendell Berry, for example, argues that real pastoral conflicts are often not city versus country but intrarural—involving local landowners, country magistrates and courts, farmhands and farm owners, and local bankers. Berry speaks directly about these very issues in his "Lake District," the fictional Port William and its Kentucky River farmlands.

Nature Writing in America

American nature writing, including its pastorals, developed distinctively New World qualities as it adapted the European nature-writing tradition to the conditions of life in North America. Nature writer and historian Peter Fritzell identifies two strands of influence that come from contrasting interests in the natural world: Aristotle's *Historia Animalium* and Saint Augustine's *Confessions*. From Aristotle we inherited a disinterested discipline of scientific observation and writing, Fritzell contends, and from Saint Augustine a very personal, autobiographical mode of writing about the metaphysical and spiritual dimensions of life. A more direct scientific influence comes from the eighteenth-century scientist Linneaus, who established a system for identifying and a framework for classifying all living things.[8]

Nineteenth-century English country cleric Gilbert White wrote the most influential text in the history of English nature writing. His *A Natural History of Selborne* (1759) brings together the contrasting strands of the scientific and pastoral traditions of nature writing. As a priest, White served the religious and metaphysical side of life in Selbourne. In writing the natural history of his village and its country environs, he was literally a pastoral voice for the spiritual inspiration and secular pleasures that could derive from life attuned to nature. Yet, armed with Linneaus, he was also a methodical scientific observer; his extensive field notes formed the basis for his rudimentary ecology of his region. White has exerted a profound and lasting influence on nature writers. Darwin, for example, took a copy of White's *Selborne* with him on H.M.S. *Beagle*, and Thoreau brought his copy to the cabin at Walden Pond. Peter Fritzell considers these two seemingly incompatible strands of nature writing that White exhibits, personal narrative and systematic observation, a paradoxical presence in American nature writing as well.[9]

Another contrast between seemingly incompatible elements that contributes to the complexity of American nature writing, especially of our pastorals, yokes a desire to escape to nature with a desire to cry out in its behalf, the warnings of a jeremiad. Thoreau's *Walden* incorporates these and many other narrative and thematic elements. Although the occasional American nature writer expresses a desire to escape from civilization altogether into a pristine, perhaps Edenic, existence outside time and history, the more common pattern involves a return to civilization from the temporary agricultural or wilderness escape, enlightened and reinvigorated. Wendell Berry tells of his escape from urban life to a permanent residence in agricultural lands, yet he does not escape from history or the social and environmental issues of his age.

The first person narrative voice has been as distinctive and significant to American nature writing and its pastorals as to our literature as a whole. As a group, these narrators have been isolated figures, frequently shunning friendships, community values, and cooperative ventures for the rugged individualism of back country isolation. They consistently question and occasionally reject society for its artificiality, corrupted values, and misguided policies, preconditions to their disillusioned retreat into nature. Standing next to such pastoral narrators from fiction as Twain's Huck Finn, Hemingway's Nick Adams, and Faulkner's Ike McCaslin are the powerful autobiographical voices of John Muir, Edward Abbey, Aldo Leopold, and Annie Dillard. American pastorals also have their favored settings, giving far more frequent attention to wilderness than farmland. They have taken the essence of the pastoral to the wild river valley or the high country, probably too often ignoring the traditional European setting of the pastoral, the rural or middle landscape of settled agriculture. While the Leopolds, Dillards, and Snyders have deliberately stalked the remote, a relatively small circle of pastoral writers, including Robert Frost, Sue Hubbell, Maxine Kumin, and Wendell Berry, have spoken for the necessity and virtue of settled agriculture.

Primarily a literature of the rural or middle landscape in England, the pastoral readily adjusted to the landscape realities of a continent of intense variety, including broad rivers and wetlands, mountain wilderness, and arid desert. Leo Marx contends that expressions of American pastoralism have held an uneasy relationship with the still-dominant progressive myth of America as a land of opportunity for material well-being and freedom from all but minimal governmental restraint on individual liberties. A contrasting and far less popular goal of establishing an Edenic agrarian society in the New World had to accommodate itself to both the rapid rise of industrialism and the need to cut and clear the land to *create* a middle landscape. As Marx has commented, to establish a pastoral landscape, "it was first necessary to transform the wilderness into a garden . . . by leveling forests, building roads, houses, farms, cities across the land."[10] Thomas Jefferson represents the first political articulation of an agrarian pastoral plan, extolling the healthful and moral virtues of a fundamentally agricultural society. Most Americans have simultaneously embraced both a progressive, industrial and a pastoral, conservationist vision of national aspirations. Widespread recognition of the fundamental incompatibility of the two visions has only recently emerged. The few nineteenth-century visionaries who understood this irreconcilability and argued the case publicly include Henry Thoreau, George Perkins Marsh (often called the father of American ecology), and John Muir (conservation activist and founder of the Sierra Club).

Many of the essays in this edition make reference to Thoreau's *Walden,* America's most significant and influential pastoral. Thoreau and the entire tradition of nature writing in America are in turn indebted to Emerson, whose essay "Nature" (1836) planted the conceptual seeds of an American nature philosophy. Emerson contributed a distinctively American conceptualization and language to the ideas of transcendence, the sublime, and the connection of nature with the soul.

In the English pastoral tradition, *Walden* extols the virtues of country life, but it also launches a pastoral attack at the corrosive impact of the values of urban capitalist society both on its trapped masses and on the surrounding countryside. Further, Thoreau leaps over the traditional agricultural landscape of the pastoral, building his cabin and laboratory in wooded land just beyond the traditional agricultural zone of the pastoral. In a gesture not uncommon to pastorals, he urges his readers to "simplify, simplify" and offers his book of country wisdom to farmers and city dwellers alike, to all who live "lives of quiet desperation." Thoreau not only establishes the essential elements of the pastoral in an American context but shifts the setting of the pastoral. By situating himself in unsettled country, he establishes "the wild" as a significant venue for American pastorals, a position from which to critique agricultural and urban policies and practices, and from which to speak persuasively for nature. His experimental life in the woods is so successful he cannot restrain himself from awakening his neighbors to the possibilities for physical and spiritual renewal that await those who follow the spirit of his advice. He argues that both individuals and cities will stagnate without steady contact with wild nature. His argument for nature comes to its conclusion with the now-famous and ecologically significant line, "in wildness is the preservation of the world."[11] Thoreau, and his mentor, Emerson, have had an enormous influence on the thought and writing of most of America's literary naturalists, including such legendary figures as Walt Whitman, John Burroughs, John Muir, Robert Frost, and Aldo Leopold.

In *The Machine in the Garden* and recent modifications of his classic thesis, Leo Marx develops the view that the clash of technology and the pastoral ideal is one of the principal metaphors of nineteenth-century American writing. Marx writes that "the ominous sounds of machines, like the sound of the steamboat bearing down upon the raft [a reference to *Huckleberry Finn*] or the train breaking in upon the idyll of Walden, reverberate endlessly in our literature."[12] Marx argues that in the most evocative and compelling works of nineteenth-century American writing, images of the pastoral ideal are violated by the sudden and disturbing presence of the machine. I suggest that this same metaphor of conflict finds even more extreme polarities of expression in the pastoral writing of the first half of the twentieth century and anticipates some of the challenges faced by writers represented in this edition.

William Faulkner's *Go Down, Moses* (1942) and its classic story "The Bear" dramatize the several stages of environmental degradation and spiritual loss. Faulkner presents the double story of young Ike McCaslin's initiation into a wilderness fraternity and the killing of the great three-toed bear Old Ben, who represents the body and spirit of the wilderness in Faulkner's south. Soon thereafter the timber rights are sold, and the seemingly endless wilderness begins to shrink visibly with each passing year. In "Delta Autumn," the final story of *Go Down, Moses,* Ike McCaslin is an old man, and the Mississippi wilderness has been reduced to a tiny, triangular residue, still large enough for small-game hunting, but no longer either a resource to be exploited or an impediment to the advance of civilization. In "Delta Autumn," nature has lost its potency to stand as a counterforce to the power of the city and its technology. The image of two equapoised forces—that is, of the machine *in,* or even the machine *and,* the garden—had become from Faulkner's perspective in the 1940s an untenable metaphor. Rather, as "Delta Autumn" implies, the inverse of the traditional terms of the pastoral was a more accurate expression of the diminishing status of nature in industrial America.

For those writers of Faulkner's generation who were attracted to nature writing and the pastoral mode (for example, Hemingway in his Nick Adams stories, *In Our Time* [1925]), enough of nature remained to satisfy personal needs, despite the encroachments of city and the machine. But any sense of equilibrium between the two forces, between industrial capitalism and wilderness and rural life, was rapidly disappearing. Like canaries in a coal mine, American nature writers have responded sooner than the body politic, often expressing through figurative language dramatic conflicts and changes that come more subtly and more slowly in society.

In American nature-writing fiction since the 1960s, one of the persistent metaphoric expressions has been of an anemic and struggling garden within a powerful machine, just the inverse of Leo Marx's original thesis regarding the dominant pastoral trope in nineteenth-century texts. This radical repositioning of the terms of the pastoral admits to a loss of balance between the two traditional forces and implies an enclave status for traditional agriculture, wilderness, and wildlife. As we approach the twenty-first century, this image of a garden within a machine becomes an increasingly accurate and compelling metaphor not only in literature but in actual conditions of world ecosystems.

For Faulkner and Hemingway, clear-cut logging and modern warfare represented the attack of the machine on the American garden, but fiction writers of the 1960s who were attracted to the pastoral identified the counterforce with automation, cybernetics, Third World exploitation, and the threat of nuclear war. For Kurt Vonnegut, Jr., Robert Coover, and Richard Brautigan, three characteristic fiction writers of the 1960s, the nature that sustained and often restored life for Emerson, Thoreau, Whitman, and such fictional characters as Huck Finn, Hemingway's Nick Adams, and Faulkner's Ike McCaslin no longer has the power to save itself from human assault. The growing awareness of environmental damage that led to the first Earth Day and the rise of an environmental movement challenged a new generation of writers to express their concerns about the loss of balance.

Vonnegut, Coover, and Brautigan express, through irony, visions of pastoral loss, if not despair, and thus anticipate some of the nature writers who receive attention in this book. Kurt Vonnegut's disturbing satire *Cat's Cradle* (1963) depicts not only the crass exploitation of Third World peasants by a pathetic dictatorship but, far worse, an "ice-nine" Armageddon that destroys their tropical island and freezes most of the life on the planet to extinction. Richard Brautigan's 1967 novel *Trout Fishing in America* also undermines the traditional pastoral hope for nature's resurgence by dramatizing the death of "Trout Fishing," a mythical wilderness spirit who dies of asphyxiation in America's polluted trout streams. Robert Coover's fiction "Morris in Chains," from *Pricksongs and Descants,* depicts in mock-serious pastoral style the imprisonment of the last of the free and independent shepherds to roam the precarious pastures of alpine America and the death of the last herd of wild mountain sheep. In their place, America has created a national park system that is computer programmed for perfect weather conditions—an electronically enhanced pastoral ideal within the insidious machine.

More recently, Peter Matthiessen's fiction, discussed in this book, works extensively with pastoralism but by implication questions the practices of his pastoral characters, most of whom have lost their way for failing to respect life and the basic ecological principles of balance and limits. As an example, his Caribbean novel *Far Tortuga* (1975) depicts the tragic end of the green turtle fishery in the Caribbean. The Grand Cayman fishermen, by contrast to Wordsworth's Cumbrian shepherds, willingly participated in the overfishing and collapse of the turtle fishery on which their island economy depended. Writing from a slightly more hopeful standpoint, Wendell Berry presents an embattled pastoral landscape in the fictional community of Port William, Kentucky, in which farm families struggle to remain good stewards of the land despite mounting pressures and temptations.

The Essays of This Volume

The writers represented in this book fall easily within the tradition of American nature writers and pastoralists, but with numerous idiosyncrasies. Nearly all these writers lived or are living in rural or backcountry locations. Most of them can be identified with fairly specific regions about which they write at length: Abbey and Krutch with the arid lands of the Southwest, Barry Lopez with the Pacific Northwest and the Arctic, Snyder with the Sierras, Leopold with Wisconsin's Sand County, Wendell Berry with northern Kentucky's Henry County, and Dillard with rural Virginia. That leaves the travel writers Matthiessen, whose settings are worldwide, and McPhee, who traverses North America but has been writing about the geology of the American west in recent books. As William Rueckert observes, Barry Lopez writes about the need to know one place well, to be a local authority, a carrier of the natural history of one's home region. From this knowledge, collected by eyes, nose, knees, and fingertips, as Berry and Lopez might put it, one can earn the opportunity to establish a dignified and honorable relationship with nature.

The work of almost every writer in this book can be connected in some way with the themes Thoreau expresses in *Walden* and his nature essays. Individual essays directly explore the connections between Thoreau's *Walden* and the work of writers as diverse as Abbey, Dillard, and Krutch. By alternating between the opposed traditions of detached, scientific observation and an intensely personal, first-person narrative that conveys autobiography, personal philosophy, and ethics, Thoreau touches on most of the themes and ideas that have concerned nature and environmental writers even in the present day.

As John Tallmadge points out in his essay, Aldo Leopold is also a Thoreauvian. His now-classic *A Sand County Almanac* serves as an exemplary pastoral guide, introducing readers to his farm and Sand County surroundings. In addition to field-guide observations about the natural life on his farm, Leopold, like Thoreau, shows his intense interest in the spiritual and ethical dimensions of existence that connect human life to nature. His major contribution comes in the form of a land ethic, a statement of land citizenship, that has been the foundation for American thinking about ecology and environment in recent years. As an explorer of diverse landscapes, both rural and wild, Leopold serves as a pastoral guide; he finds startling truths in nature, as John Tallmadge observes, and writes to inform human society for its own good and the good of nature. Like all pastoral guides, he urges his readers not just to know nature better but to know or "read" themselves better.

James McClintock observes that Annie Dillard "uses nature as a touchstone for spiritual insight," a comment that could be applied to Wendell Berry and Gary Snyder. In the tradition of pastoral guides, Dillard is a walker, a stalker, a seer, inviting all who will join in her nature pilgrimage to heightened awareness. Patrick Murphy finds a common strand of spirituality running through the poetry of Snyder and Berry, even though Snyder embraces the Buddhist tradition, and Berry the Christian.

Most of the nature writers find occasion to critique human civilization and in particular the practices of American industrial society. Their critiques are usually linked to a vision of harmony with nature. Barry Lopez is a tireless observer of the interplay between human and biocentric worlds; in his view humans need to relearn such ancient knowledge as how to dance with herons, how to dance with nature, and how to fulfill our natural responsibilities. Lopez argues that we humans need to radically adjust our human greed to coincide with the laws and limits of the ecosystem. Diane Wakoski thinks Abbey (and with him perhaps Snyder) is on the inhumanist or biocentric edge of this group of writers, because he considers humans interesting but ultimately irrelevant to evolution.

As one might expect, there is a loosely knit kinship of contemporary nature writers. They are well aware of each other's work and find occasions to enter into dialogue. In this book, Wendell Berry writes about Abbey, and Abbey describes his first meeting and interview with

Joseph Wood Krutch. Even though they represented very different traditions, Abbey respected Krutch's humanism and speaks of the "planetary significance" of his words. Most of the authors of essays included here are in contact with other critics and students of nature writing, and a number belong to the recently formed Association for the Study of Literature and the Environment. Its affiliated publication, *The American Nature Writing Newsletter,* publishes reviews, essays, and information about activities relating to the study of writing on nature and the environment.

Because of the impact of growing environmental awareness, we should expect a continuing, rigorous dialogue about the future of nature writing. Important contributions to this dialogue will come from ecofeminism (which has established linkages between deep ecology and the feminist movement) and the emerging ecocriticism (see the Afterword). Suggesting the nature of this dialogue, Glenn Love has argued that nature writing, particularly the pastoral, is so intricately woven through the paternalistic, liberal tradition that it will be difficult to make the necessary transformation from "ego to eco-consciousness."[13] A few critical voices have argued that nature writing has tended toward a consciousness excessively bifurcated (city versus country) and preoccupied with landscape zones or with the pursuit of individual enlightenment, often to the exclusion of pressing ecological issues. The weakness in its paternalistic and anthropocentric orientation, some critics have argued, is a reluctance to consider the emerging biocentrism, with its widely embracing recognition of diverse biotic, including human, needs and rights. Nature writing, including the pastoral, will have to shake excessive identification with "old boy" consciousness and the urbanite's desire for rustic retreat. It will have to find new sensitivities and dimensions of understanding to be a fully comprehensive modality for the future expression of eco-consciousness. This process of challenge and change is occurring. As a more diverse constituency of nature writers (including pastoralists) achieves wide recognition, the genre will become a more comprehensive reflection of the environmental issues and concerns of this crucial period.

This collection of essays demonstrates the adaptability of nature writing, including its pastoral mode, to the interests of women as well as men writers, to wilderness and arid-land writers as well as to writers of farm and rural life. The pastoral continues to attract writers and readers because it is a modality flexible enough to embrace explorations into the nature of society, flexible enough to embrace ideas for a sustainable balance between human activities and the biosphere. Lawrence Buell argues the importance of placing the pastoral in a postcolonial context, which highlights its "ideological multivalence" and its presence in minority writing. Buell stresses that American pastorals run a very wide range from recessive to radical, from retreat (often as reenactment of pioneer experience) to alienated assault on mainstream values, and that they are thus well suited for our contemporary cultural dialogue.[14] Glenn Love sees the continuing appeal of the pastoral as a "testament to our instinctive or mythic sense of ourselves as creatures of natural origins ... who must return periodically to the earth for the rootholds of sanity somehow denied us by civilization."[15] In addition to this timelessly appealing dimension of the pastoral, which reflects the particular crisis of modern industrial societies, there is great need for nature writing that probes challenging subjects and draws from new ideas, including ecofeminism, contemporary literary theory and ecocriticism, deep ecology, green politics, and postcolonialism.

NOTES

1. Leo Marx, "Pastoralism in America," in *Ideology in Classic American Literature,* ed. Sacvan Bercovitch and Myra Jehlen (New York: Cambridge University Press, 1987), 43.

2. Ibid., 43.

3. See, in this book, Patrick Murphy's article on Wendell Berry and Gary Snyder, and John Cooley's article on Peter Matthiessen. Pastoral decay and loss of Eden are themes explored in Richard Brautigan's *Trout Fishing in America* (New York: Delta Publishing Co., 1969), Robert Coover's "Morris in Chains," in *Pricksongs and Descants* (New York: New American Library, 1969), and Kurt Vonnegut, Jr.'s *Cat's Cradle* (New York: Dell Publishing Co., 1963).

4. Harold Tolliver, *Pastoral Forms and Attitudes* (Berkeley: University of California Press, 1971), 89.

5. William Empson, *Some Versions of Pastoral* (New York: New Directions, 1968), 22.

6. Raymond Williams, *The Country and the City* (London: Oxford University Press, 1974), 302.

7. Roger Sales, *English Literature and History 1780–1830* (New York: St. Martin's Press, 1984), 15.

8. See, for example, his *Systema Naturae* (1735) and *Species Plantarum* (1753).

9. For a comprehensive study of the history and characteristics of nature writing, see Peter A. Fritzell, *Nature Writing and America* (Ames: Iowa University Press, 1990).

10. Leo Marx, "Pastoralism in America," 49.

11. *Walden; or, Life in the Woods* (New York: Rinehart & Co., 1948), 265.

12. Leo Marx, *The Machine in the Garden: Technology and the Pastoral Ideal in America* (New York: Oxford University Press, 1964), 27. See also Marx's *The Pilot and the Passenger: Essays on Literature, Technology and Culture in the United States* (New York: Oxford University Press, 1988).

13. Glenn A. Love, "Revaluing Nature: Toward an Ecological Criticism," *Western American Literature* 25, no. 3 (1990): 207.

14. "American Pastoral Ideology Reappraised," *Norton Anthology of American Literature* (New York: W. W. Norton, 1992), 270.

15. Love, "Revaluing Nature," 207.

Edward Abbey
1927–89

"All men are brothers, we like to say, half-wishing sometimes in secret it were not true. But perhaps it is true. And is the evolutionary line from protozoan to Spinoza any less certain? That also may be true. We may be obliged, therefore, to spread the news, painful and bitter though it may be for some to hear, that all livings things on earth are kindred."

—Edward Abbey, *Desert Solitaire*

Edward Abbey was born in Home, Pennsylvania, on January 29, 1927. He received his college education at the University of New Mexico, graduating with a B.A. in 1951 and an M.A. in 1956. In addition to his distinguished literary career (thirty books and numerous articles), Abbey held a remarkable variety of jobs; he was a fire watcher and a park ranger for the National Park Service, a bus driver, a social worker, a cowboy, and a movie consultant. Politically, Abbey considered himself an agrarian anarchist, and theologically he was closest to the religion of the Piutes.

Abbey refused to consider himself a nature writer or an environmentalist, even though the wilderness and open spaces of the Southwest form the settings for nearly all his books. His writing shows a passionate love for wilderness and a distrust of human society and its institutions. Among contemporary nature writers, he is the most critical of American society.

Abbey's cultural criticism cuts to the heart of American assumptions about progress. With assertions like "we are slaves of . . . an expand-or-expire agro-industrial empire," Abbey provoked to disturb the peace. In addition to his skillful satire, his voice sometimes turned to open frustration and anger, and he sought desert solitude from the society he barely tolerated.

Like Thoreau, Abbey critiqued American civilization and found it sorely lacking. It has honored comfort and security over the risk and challenge of living directly and openly. He argued that life needs to be restored to "the firm reality of mother earth."

As Wendell Berry observes, Abbey was "a great irreverencer of sacred cows." He refused to speak for or be identified with any group or movement, including the environmental movement. Despite many efforts to pigeonhole him, Edward Abbey steadfastly refused to view the world in any way but his own.

A Few Words in Favor of Edward Abbey

Wendell Berry

Reading through a sizeable gathering of reviews of Edward Abbey's books, as I have lately done, one becomes increasingly aware of the extent to which this writer is seen as a problem by people who are, or who think they are, on his side. The problem, evidently, is that he will not stay in line. No sooner has a label been stuck to his back by a somewhat hesitant well-wisher than he runs beneath a low limb and scrapes it off. To the consternation of the "committed" reviewer, he is not a conservationist or an environmentalist or a boxable *-ist* of any other kind; he keeps on showing up as Edward Abbey, a horse of another color, and one that requires care to appreciate.

He is a problem, apparently, even to some of his defenders, who have an uncontrollable itch to apologize for him: "Well, he did *say* that. But we mustn't take him altogether seriously. He is only trying to shock us into paying attention." Don't we all remember, from our freshman English class, how important it is to get the reader's attention?

Some environmentalist reviewers see Mr. Abbey as a direct threat to their cause. They see him as embarrassingly prejudiced or radical or unruly. Not a typical review, but representative of a certain kind of feeling about Edward Abbey, was Dennis Drabelle's attack on *Down the River* in *The Nation*, 1 May 1982. Mr. Drabelle accuses Mr. Abbey of elitism, iconoclasm, arrogance, and xenophobia; he finds that Mr. Abbey's "immense popularity among environmentalists is puzzling"; and observes that "many of his attitudes give aid and comfort to the enemies of conservation."

Edward Abbey, of course, is a mortal requiring criticism, and I would not attempt to argue otherwise. He undoubtedly has some of the faults he has been accused of having, and maybe some others that have not been discovered yet. What I *would* argue is that attacks on him such as that of Mr. Drabelle are based on misreading, and that the misreading is based on the assumption that Mr. Abbey is both a lesser man and a lesser writer than he is in fact.

Mr. Drabelle and others like him assume that Mr. Abbey is an environmentalist—hence, that they, as other environmentalists, have a right to expect him to perform as their tool. They further assume that, if he does not so perform, they have a proprietary right to complain. They would like, in effect, to brand him an outcast and an enemy of their movement, and to enforce their judgment against him by warning people away from his books. Why should environmentalists want to read a writer whose immense popularity among them is puzzling?

Such assumptions, I think, rest on yet another that is more important and more needful of attention: the assumption that our environmental problems are the result of bad policies, bad political decisions, and that, therefore, our salvation lies in winning unbelievers to the right political side. If all those assumptions were true, then I suppose that the objections of Mr. Drabelle would be sustainable; Mr. Abbey's obstreperous traits would be as unsuitable in him as in any other political lobbyist. Those assumptions, however, are false.

Mr. Abbey is not an environmentalist. He is, certainly, a defender of some things that environmentalists defend, but he does not write merely in defense of what we call "the environment." Our environmental problems, moreover, are not, at root, political; they are cultural. As Edward Abbey knows and has been telling us, our country is not being destroyed by bad politics; it is being destroyed by a bad way of life. Bad politics is merely another result. To see that the problem is far more than political is to return to reality, and a look at reality permits us to see, for example, what Mr. Abbey's alleged xenophobia amounts to.

The instance of xenophobia cited by Mr. Drabelle occurs on page 17 of *Down the River,* where Mr. Abbey proposes that our Mexican border should be closed to immigration. If we permit unlimited immigration, he says, before long

> the social, political, economic life of the United States will be reduced to the level of life in Juarez. Guadalajara. Mexico City. San Salvador.

Haiti. India. To a common peneplain of overcrowding, squalor, misery, oppression, torture, and hate.

That is certainly not a liberal statement. It expresses "contempt for other societies," just as Mr. Drabelle says it does. It is, moreover, a fine example of the exuberantly opinionated Abbey sentence that raises the hackles of readers like Mr. Drabelle—as it is probably intended to do. But before we dismiss it for its tone of "churlish hauteur," we had better ask if there is any truth in it.

And there is some truth in it. As the context plainly shows, this sentence is saying something just as critical of ourselves as of the other countries mentioned. Whatever the justice of the "contempt for other societies," the contempt for the society of the United States, which is made explicit in the next paragraph, is fearfully just.

We are slaves in the sense that we depend for our daily survival upon an expand-or-expire agro-industrial empire—a crackpot machine—that the specialists cannot comprehend and the managers cannot manage. Which is, furthermore, devouring world resources at an exponential rate. We are, most of us, dependent employees.

This statement is daily verified by the daily news. And its truth exposes the ruthless paradox of Mexican immigration: Mexicans cross the border because our way of life is extravagant; we have no place for them, or will not for very long. A generous immigration policy would be contradicted by our fundamentally ungenerous way of life. Mr. Abbey assumes that before talking about generosity, we must talk about carrying capacity, and he is correct. The ability to be generous is finally limited by the availability of supplies.

The next question, then, must be: If he is going to write about immigration, why does he not do it in a sober, informed, logical manner? The answer, I am afraid, will not suit some advocates of sobriety, information, and logic: he *can* write in a sober, informed, logical manner—if he *wants* to. And why does he sometimes not want to? Because it is not in his character to want to all the time. With Mr. Abbey character is given, or it takes, a certain precedence, and that precedence makes him a writer and a man of a different kind, and probably a better kind, than the practitioner of mere sobriety, information, and logic.

In classifying Mr. Abbey as an environmentalist, Mr. Drabelle is implicitly requiring him to be sober, informed, and logical. And there is nothing illogical about Mr. Drabelle's discomfort when his call for an environmentalist is answered by a man of character, somewhat unruly, who apparently did not know that an environmentalist was expected. That, I think, is Mr. Abbey's problem with many of his detractors. He is advertised as an environmentalist. They *want* him to be an environmentalist. And who shows up but this *character,* who writes beautifully some of the time, who argues some of the time with great eloquence and power, but who some of the time offers opinions that appear to be only his own uncertified prejudices, and who some of the time, and even in the midst of serious discussion, makes *jokes.*

If Mr. Abbey is not an environmentalist, what is he? He is, I think, at least in the essays, an autobiographer. He may be writing on one or another of what are not called environmental issues, but he remains Edward Abbey, speaking as and for himself, fighting, literally, for dear life. This is important, for if he is writing as an autobiographer, he *cannot* be writing as an environmentalist—or as a specialist of any other kind. As an autobiographer, his work is self-defense. As a conservationist, he is working to conserve himself as a human being. But this is self-defense and self-conservation of the largest and noblest kind, for Mr. Abbey understands that to defend and conserve oneself as a human being in the fullest, truest sense, one must defend and conserve many others and much else. What would be the hope of being personally whole in a dismembered society; or personally healthy in a land scalped, scraped, eroded, and poisoned; or personally free in a land entirely controlled by the government; or personally enlightened in an age illuminated only by TV? Edward Abbey is fighting on a much broader front than that of any "movement." He is fighting for the survival not only of nature but of *human* nature, of culture, as only our heritage of works and hopes can define it. He is, in short, a traditionalist—as he has said himself, expecting, perhaps, not to be believed.

Here the example of Thoreau becomes pertinent. My essay may seem on the verge of becoming very conventional now, for one of the strongest of contemporary conventions is that of comparing every writer who has been as far out of the house as the mailbox to Thoreau. But I do not intend to say that Mr. Abbey writes like Thoreau, for I do not think he does, but only that their cases are similar. Thoreau has been adopted by the American environmental movement as a figurehead;

he is customarily quoted and invoked as if he were in some simple way a forerunner of environmentalism. This is possible, obviously, only because Thoreau has been dead since 1862. Thoreau was an environmentalist in exactly the same sense that Edward Abbey is: he was for some things that environmentalists are for. And in his own time he was just as much an embarrassment to movements, just as uncongenial to the group spirit, as Edward Abbey is, and for the same reasons: he was working as an autobiographer, and his great effort was to conserve himself as a human being in the best and fullest sense. As a political activist, he was a poor excuse. What was the political value of his forlorn, solitary taxpayer's revolt against the Mexican War? What was politic about his defense of John Brown, or his insistence that abolitionists should free the *wage* slaves of Massachusetts? Who could trust the diplomacy of a man who would pray in his poem "My Prayer":

Great God, I ask thee for no other pelf
Than that I may not disappoint myself;
. . . .
And next in value, which thy kindness lends,
That I may greatly disappoint my friends.

The point, evidently, is that if we want the human enterprise to be defended, we must reconcile ourselves to the likelihood that it can be defended only by human beings. This, of course, entails an enormous job of criticism: an endless judging and sorting of the qualities of human beings and of their contributions to the human enterprise. But the size and urgency of this job of criticism should warn us to be extremely wary of specializing the grounds of judgment. To judge a book by Edward Abbey by the standard of the immediate political aims of the environmentalist movement is not only grossly unfair to Mr. Abbey but a serious disservice to the movement itself.

The trouble, then, with Mr. Abbey—a trouble, I confess, that I am disposed to like—is that he speaks insistently as himself. In any piece of his, we are apt to have to deal with all of him, caprices and prejudices included. He does not simply submit to our criticism, as does any author who publishes, but virtually demands it. And so his defenders, it seems to me, are obliged to take him seriously, to assume that he generally means what he says, and, instead of apologizing for him, to acknowledge that he is not always right or always fair. He is *not*, of course. Who is? For me, part of the experience of reading him has always been, at certain points, that of arguing with him.

My defense of him begins with the fact that I want him to argue with, as I want to argue with Thoreau, another writer full of cranky opinions and strong feelings. If we value these men and their work, we are compelled to acknowledge that such writers are not made by tailoring to the requirements, and trimming to the tastes, of any and all. They submit to standards raised, though not made, by themselves. We, with our standards, must take them as they come, defend ourselves against them if we can, agree with them if we must. If we want to avail ourselves of the considerable usefulness and the considerable pleasure of Edward Abbey, we will have to like him as he is. If we cannot like him as he is, then we will have to ignore him, if we can. My own notion is that he is going to become harder to ignore, and for good reasons— not the least being that the military-industrial state is working as hard as it can to prove him right.

It seems virtually certain that no reader can read much of Mr. Abbey without finding some insult to something that he or she approves of. Mr. Abbey is very hard, for instance, on "movements"—the more solemn and sacred they are, the more thay tempt his ridicule. He is a great irreverancer of sacred cows. There is, I believe, not one sacred cow of the sizeable herd still on the range that he has left ungoosed. He makes his rounds as unerringly as the local artificial inseminator. This is one of his leitmotifs. He gets around to them all. These are glancing blows, mainly on the run, with a weapon no more lethal than his middle finger. The following is a fairly typical example, found in his introduction to *Down the River*: "The essays in *Down the River* are meant to serve as antidotes to despair. Despair leads to boredom, electronic games, computer hacking, poetry and other bad habits." That example is appropriate here because it passingly gooses one of my own sacred cows: poetry. I am inclined to be tickled rather than bothered by Mr. Abbey's way with consecrated bovines, and this instance does not stop me long. I do pause, nevertheless, to think that *I*, anyhow, would not equate poetry with electronic pastimes. But if one is proposing to take Mr. Abbey seriously, one finally must stop and deal with such matters. Am I, then, a defender of "poetry"? The answer, inevitably, is no; I am a defender of some poems. Any human product or activity that humans defend as a category becomes, by that fact, a sacred cow—in need, by the same fact, of an occasional goosing, an activity, therefore, that arguably serves the public good.

Some instances are funnier than others, and readers will certainly disagree as to the funniness of any given instance. But whatever one's opinion, in particular or in general, of Mr. Abbey's blasphemies against sacred cows, one should be wary of the assumption that they are merely humorous, or (as has been suggested) merely "image-making" stunts calculated to sell articles to magazines. They are, I think, gestures or reflexes of his independence—his refusal to speak as a spokesman or a property of any group or movement, however righteous. This keeps the real dimension and gravity of our problem visible to him, and keeps him from falling for easy answers. You never hear Mr. Abbey proposing that the fulfillment of this or that public program or the achievement of the aims of this or that movement, or the "liberation" of this or that group will save us. The absence, in him, of such propositions is one of his qualities, and it is a welcome relief.

The funniest and the best of these assaults are the several that are launched head-on against the most exalted of all the modern sacred cows: the self. Mr. Abbey's most endearing virtue as an autobiographer is his ability to stand aside from himself and recount his most outrageous self-embarrassing goof-ups with a bemused and gleeful curiosity, as if they were the accomplishments not merely of somebody else but of an altogether different kind of creature. I envy him that. It is, of course, a high accomplishment. How absurd we humans, in fact, are! How misapplied is our self-admiration—as we can readily see by observing other self-admiring humans! How richly just and healthful is self-ridicule! And yet how few of us are capable of it. I certainly do find it hard. My own goof-ups seem to me to have received merciless publicity when my wife has found out about them.

Because he is so humorous and unflinching an autobiographer, he knows better than to be uncritical about anything human. That is why he holds sacred cows in no reverence. And it is at least partly why his reverence for nature is authentic; he does not go to nature to seek himself or flatter himself, or speak of nature in order to display his sensitivity. He is understandably reluctant to reveal himself as a religious man, but the fact occasionally appears plainly enough: "It seems clear at last that our love for the natural world—Nature—is the only means by which we can requite God's obvious love for it!"

The richest brief example of Abbey humor that I remember is his epigram on "gun control" in his essay "The Right to Arms." "If guns are outlawed," he says, "only the government will have guns." That sentence, of course, is a parody of the "gun lobby" bumpersticker: "If guns are outlawed, only outlaws will have guns." It seems at first only another example of sacred cow goosing—howbeit an unusually clever one, for it gooses both sacred cows involved in this conflict: the idea that because guns are used in murders, they should be "controlled" by the government; and the idea that the Second Amendment to the Bill of Rights confers a liberty that is merely personal. Mr. Abbey's sentence, masquerading as an instance of his well-known "iconoclasm," slices cleanly through the distractions of the controversy to the historical and constitutional roots of the issue. The sentence is, in fact, an excellent gloss on the word *militia* in the Second Amendment. And so what might appear at first to be merely an "iconoclastic" joke at the expense of two public factions becomes, on examination, the expression of a respectable political fear and an honorable political philosophy, a statement that the authors of our constitution would have recognized and welcomed. The epigram is thus a product of wit of the highest order, richer than the excellent little essay that contains it.

Humor, in Mr. Abbey's work, is a function of his outrage, and is therefore always necessary to necessity. Without his humor, his outrage would be intolerable—as, without his outrage, his humor would often be shallow or self-exploitive. The indispensable work of his humor, as I see it, is that it keeps bringing the whole man into the job of work. Often, the humor is not so much a property of the argument at hand as it is a property of the stance from which the argument issues.

Mr. Abbey writes as a man who has taken a stand. He is an *interested* writer. This exposes him to the charge of being prejudiced, and prejudiced he certainly is. He is prejudiced against tyranny over both humanity and nature. He is prejudiced in favor of democracy and freedom. He is prejudiced in favor of an equitable and settled domestic life. He is prejudiced in favor of the wild creatures and their wild habitats. He is prejudiced in favor of charitable relations between humanity and nature. He has other prejudices, too, but I believe that those are the main ones. All of his prejudices, major and minor, identify him as he is, not as any reader would have him be. Because he speaks as himself, he does not represent any group, but he stands for all of us.

He is, I think, one of the great defenders of the idea of property. His novel *Fire on the Mountain* is a moving, eloquent statement on behalf of the personal proprietorship of land—*proper* property. And this espousal of the cause of the private landowners, the small farmers and small ranchers, is evident throughout his work. But his advocacy of that kind of property is balanced by his advocacy of another kind: public property, not as "government land," but as wild land, wild property, which, belonging to nobody, belongs to everybody, including the wild creatures native to it. He understands better than anyone I know the likelihood that one kind of property is not safe without the other. He understands, that is, the natural enmity between tyranny and wilderness. "Robin Hood, not King Arthur," he said, "is the real hero of English legend."

You cannot lose your land and remain free; if you keep your land, you cannot be enslaved. That is an old feeling that began to work its way toward public principle in our country at about the time of the Stamp Act. Mr. Abbey inherits it fully. He understands it both consciously and instinctively. It is this and not nature love, I think, that is the real motive of his outrage. His great fear is the fear of dispossession.

But his interest is not just in *landed* property. His enterprise is the defense of all that properly belongs to us, including all those thoughts and works and hopes that we inherit from our culture. His work abounds in anti-intellectual jokes—he is not going to run with *that* pack, either—but no one can read him attentively without realizing that he has read well and widely. His love for Bach is virtually a theme of his work. His outrage often vents itself in outrageousness, and yet it is the outrage of a cultivated man—that is why it is valuable to us, and why it is interesting.

He is a cultivated man. And he is a splendid writer. Readers who allow themselves to be distracted by his jokes at their or our or his own expense cheat themselves out of a treasure. The xenophobic remark that so angers Mr. Drabelle, for example, occurs in an essay, "Down the River with Henry Thoreau," which is an excellent piece of writing—entertaining, funny some of the time, aboundingly alive and alert, variously interesting, diversely instructive. The river is the Green, in Utah: the occasion is a boat trip by Mr. Abbey and five of his friends in November 1980. During the trip he read *Walden* for the first time since his school days. This subjection of a human product to "the prehuman sanity of the desert" is characteristic of Mr. Abbey's work, the result of one of his soundest instincts. His account of the trip is, at once, a travelogue, a descriptive catalog of natural sights and wonders, and a literary essay. It is an essay in the pure, literal sense: a trial. Mr. Abbey tries himself against Thoreau, and Thoreau against himself; he tries himself and Thoreau against the river; he tries himself and Thoreau and the river against modern times, and vice versa. The essay looks almost capriciously informal; only a highly accomplished and knowledgeable writer would have been capable of it. It is, among all else, a fine literary essay—such a reading of *Walden* as Thoreau would have wanted, not by the faceless automaton of current academic "scholarship," but by a man outdoors, whose character is in every sentence he writes.

I do not know that that essay, good as it is, is outstanding among the many that Mr. Abbey has written. I chose to speak of it because Mr. Drabelle chose to speak of it, and because I think it represents its author well enough. It exhibits one of his paramount virtues as a writer, a virtue paramount in every writer who has it: he is always interesting. I have read, I believe, all of his books except one, and I do not remember being bored by any of them. One reason for this is the great speed and activity of his pages; a page of his, picked at random, is likely, I believe, to have an unusual number of changes of subject, and to cover an unusual amount of ground. Another reason is that he does not oversimplify either himself or, despite his predilection for one-liners, his subject. Another reason is his humor, the various forms of which keep breaking through the surface in unexpected places, like wet-weather springs.

But the quality in him that I most prize, the one that removes him from the company of the writers I respect and puts him in the company, the smaller company, of the writers I love, is that he sees the gravity, the great danger, of the predicament we are now in, he tells it unswervingly, and he defends unflinchingly the heritage and the qualities that may preserve us. I read him, that is to say, for consolation, for the comfort of being told the truth. There is no longer any honest way to deny that a way of living that our leaders continue to praise is destroying all that our country is and all the best that it means. We are living even now among punishments and ruins. For those who know this, Edward Abbey's books will remain an indispensable solace. His essays, and his novels too, are "antidotes to despair." For those who think that a few more laws will enable us to go on safely as we are going, Abbey's books will remain—and good for him!—a pain in the neck.

Edward Abbey: Joining the Visionary "Inhumanists"

Diane Wakoski

> I think I could turn and live with animals, they are so placid and self-contain'd,
> I stand and look at them long and long.
>
> They do not sweat and whine about their condition,
> They do not lie awake in the dark and weep for their sins,
> They do not make me sick discussing their duty to God,
> Not one is dissatisfied, not one is demented with the mania of owning things,
> Not one kneels to another, nor to his kind that lived thousands of years ago,
> Not one is respectable or unhappy over the whole earth.
>
> —pt. 32, Walt Whitman, "Song of Myself"

The term *inhumanist* is one coined by Robinson Jeffers to mean a person who rejects the philosophical tradition of the humanists, that tradition created in classical Greece which has dominated Western civilization, which sees all human endeavor as the central purpose of life, and in fact sees all life as having importance relative to (and less than) human life and activity. But just as the American culture has spawned a new poetry, which really sees no need for classical metrical conventions and practices to make verse, it has also spawned a tradition growing out of both scientific awareness and rebellion against European civilization's domination, that says perhaps the human end is not really so important as we think. In Jeffers's poetry, much of it focused on the natural landscape of Carmel and Big Sur in central California, where he lived, the message is not only a cosmic one—the earth is but a small part of the universe, and humans are such an infinitely small part of the possibilities of life—but also a neo-Darwinian one in which humankind is a sort of evolutionary mistake, having turned into a murdering, raping, torturing, ravaging species which will certainly destroy itself while other life in the universe, perhaps even on the planet, will live on.

Even though the passage quoted above shows that possibility in Whitman, he is for the most part a true humanist who simply chooses to believe that all mankind *could* be filled with love if it would, and that slavery, war, and other ignominies will be wiped away when his bigger vision is obtained. Still, all the seeds of "inhumanism" are planted there in that vision which does see man as pillaging rather than loving, and does idealize animals for at least not having the worst human vices. Jeffers, on the other hand, sees the human drama playing itself out, doomed and fascinating in its fated self-destruction. When Jeffers's editors at Random House in the 1940s finally realized what he was saying and were confronted by his antiwar politics in very specific terms, lumping Roosevelt, great American hero, along with Hitler and Mussolini, they not only censored some of the poems in *The Double Axe* (containing the long poem with one section entitled "The Inhumanist") but also wrote an editorial note, placed in the front of the book, disclaiming any of the ideas expressed in the book. The fear, not just patriotic or chauvinistic, of expressing a feeling that the human race was not the most important thing we could know shocked everyone. This was not politics. It was an undermining of civilization.

Racing along under the surface of all of Ed Abbey's writing is that fiery "inhumanist" philosophy. It makes him love the desert above all things and fuels his equal desire to be in the wilderness anywhere, to explore and understand and simply be with the nonman, the ahuman, world. But pumping just as strong as a heart inside him is the tradition, his education, and his feeling that we must believe in human civilization, must try to save it, about which Abbey equivocates often to try to understand these contradictory urges in himself. In *Desert Solitaire,* near the end of Abbey's sojourn in Arches National Park as a ranger, on Labor Day weekend, a stranger who signs himself J. Prometheus Birdsong keeps him up all night discussing philosophy and what his real position is vis-à-vis the humanist issue. They decide that his only reason for arguing the humanist position at all is that he is human and cannot quite face the possibility that he is of no importance whatsoever. And in another discussion with his friend and fellow camper Ralph Newcombe, Abbey decides that he is not an atheist but an "earthiest." "Be true to the earth" is his motto.[1] Yet it is this struggle with the human need to survive, triumph, and continue in society and civilization, along with his feeling that humans are irrelevant to the cosmos, that makes Abbey's writing so rich. We are not being palmed off with nature worship, nor are we being forced to see anything but the reality of twentieth-century man, who has immense resources and chronically uses them badly. At one point in *Desert Solitaire,* Abbey makes a distinction between "civilization" and "culture." Even though most of his examples are frivolous, the distinction is eminent in the argument for Abbey's aesthetic, which I think is neither Whitman's longing ("I stand and look at them and long and long") for the world of humankind to be as free and pure of destructive vices, nor Jeffers's cynical belief that man is simply a mistake in evolution which the very process of evolution will soon make right and that war will simply destroy the planet.

Abbey is ultimately both politician and poet, spending half of his year in the wilderness, half in civilization, working for the Park Service and giving little programs of "revolution," as he calls them, by which we could set ourselves back on a constructive course. At the same time, he sees more and more the desert as a symbol for some ineffable greatness (God?) that it is in man's power to approach, perhaps contain. He says, "I am convinced now that the desert has no heart, that it presents a riddle which has no answer, and that the riddle itself is an illusion created by some limitation or exaggeration of the displaced human consciousness" (273). And yet apparently for Abbey, the illusion is also that it does not matter whether one attains the answer, but whether one is allowed to continue the pursuit for it.

Strangely, Abbey is no prophet of doom. Like the desert, he seems to offer philosophies which do not bring final answers but lead one to other questions. The desert which is his passion is loved because it is one of the last things which no one could want to own. And even when it is temporarily co-opted for uranium ore or other precious minerals, it is always soon wasted again and finally left to the Abbeys of the world, those who do not want to own or exploit but only to be. He is eloquent on the need for wilderness on this planet. We need places where no one could choose to be, which because of that will be underdeveloped and thus symbols of freedom.

> The knowledge that refuge is available, when and if needed, makes the silent inferno of the desert more easily available, when and if needed, makes the silent inferno of the desert more easily bearable. Mountains complement desert as desert complements city, as wilderness complements and completes civilization.

> A man could be a lover and defender of the wilderness without ever in his lifetime leaving the boundaries of asphalt, powerlines, and right-angled surfaces. We need wilderness whether or not we ever set foot in it. We need a refuge even though we may never need to go there. I may never in my life get to Alaska, for example, fur I am grateful that it's there. We need the possibility of escape as surely as we need hope; without it the life of the cities would drive all men into crime or drugs or psychoanalysis. (*Desert Solitaire,* 148–49)

This vision of Abbey's is often co-opted for trendy and fashionable uses, rather than used salvagingly as it might be. For Abbey, like Whitman and Jeffers, is trying to find a way to understand his own humanness and failures while not dooming the entire human race. But the inherent paradox in this is unavoidable. In a rhapsodic passage near the beginning of *Desert Solitaire,* he describes killing a rabbit with a stone, just for the joy of being in the wilderness and being able to do it, not for meat, and not because he is actually a hunter. He is trying to feel himself a part of the landscape. Yet, earlier that same week he has had the problem of what to do about mice in his trailer. They attract snakes, and he does not want to live with rattlesnakes. He does not want to kill the mice. He does not want either to have to kill the Faded Midget, a little horned rattlesnake which has come to live under the steps. Nature solves the problem for him, in one of the most charming parts of *Desert Solitaire,* when Abbey tells the story of his living with a bull snake for a few weeks, in April when it is still cold. The snake loves the warm trailer and often curls itself around Abbey's waist, inside his shirt. It drives away both the Faded Midget from the doorstep and the mice. But Abbey deliberately sets up his wish not to kill—his humanist self—for the reader; then he takes his walk into the desert, where he savagely kills the rabbit for no reason at all.

> For a moment I am shocked at my deed; I stare at the quiet rabbit, his glazed eyes, his blood drying in the dust. Something vital is lacking. But shock is succeeded by a mild elation. Leaving me victim to the vultures and maggots, who will appreciate him more than I could—the flesh is probably infected with tularemia—I continue my walk with a new, augmented cheerfulness which is hard to understand but unmistakable. What the rabbit has lost in energy and spirit seems added, by processes too subtle to fathom, to my own soul. I

try but cannot feel any sense of guilt. I examine my soul: white as snow. Check my hands: not a trace of blood. No longer do I feel so isolated from the sparse and furtive life around me, a stranger from another world. I have entered into this one. We are kindred all of us, killer and victim, predator and prey, me and the sly coyote, the soaring buzzard, the elegant gopher snake, the trembling cottontail, the foul worms that feed on our entrails, all of them, all of us. Long live diversity, long live the earth! (38–39)

What finally, then, is Abbey's vision?

Yes, we can say the Dionysian, in which we understand all of life to be part of a cycle, death as much a part of reality as birth, and death required before rebirth can occur. But what explains Abbey's tirades against "culture" rather than "civilization," his anger against the motor vehicle, his hatred of all the tourists who come to the park? This seems to be different from Jeffers's "inhumanism" when looked at entirely; there is no real conviction here that all humanity is a mistake. If there is any political message constantly in Abbey's writings, it seems to be that we have overpopulated the world, not that mankind is bad, only that certain humans are, and that in large numbers humankind is trouble.

Is it specious to conclude that Abbey, like his desert, presents riddles which have not answers? Perhaps. Yet maybe that is part of the appeal of his work in this time when we are aware of a very probably approaching nuclear holocaust. Perhaps his lack of doctrine or dogma is reassuring in itself.

> The desert says nothing. Completely passive, acted upon but never acting, the desert lies there like the bare skeleton of Being, spare, sparse, austere, utterly worthless, inviting not love but contemplation. In its simplicity and order it suggests the classical, except that the desert is a realm beyond the human and in the classical view only the human is regarded as significant or even recognized as real. (*Desert Solitaire*, 270)

This meditative line, even though not the same as Jeffers's conclusions, is an "inhumanist" speculation. Certainly, it is what is beyond the human in the desert and wilderness which draws Abbey. And all of his readers must thank him, as we thank Whitman and Jeffers, for giving us some respite from our own resolutely self-centered, and probably destructive, humanism.

NOTE

1. Edward Abbey, *Desert Solitaire* (New York: Ballantine Books, 1968), 208.

Wendell Berry
1934–

> An enduring agriculture must never cease to consider and respect and preserve wildness. The farm can exist only within the wilderness of mystery and natural force. And if the farm is to last and remain in health, the wilderness must survive within the farm.
>
> —Wendell Berry, "The Body and the Earth,"
> in *The Unsettling of America*

Unconsciously perhaps from the beginning, and more and more consciously during the last sixteen or seventeen years, my work has been motivated by a desire to make myself responsible at home both in this world and in my native and chosen place. (*Recollected Essays*)

These are the words of writer and poet Wendell Berry, a man whose literary works celebrate his commitment to the land. In his novels, essays, and poems, he writes about the middle landscape, the garden, and those who occupy it. Taking his place between the urban dweller on one side and unbounded wilderness on the other, Berry speaks, as a farmer, for the farm, and for a vision of man and nature working in harmony. His goal is, in part, to bring health to the environment through responsible agriculture. For Berry, the small family farm, worked in harmony with nature, is the only permanent way to restore our depleted earth. We need to come back to the soil, he tells us, and to rediscover a sense of place—a sense of community and commitment to the land. In this book, Jack Hicks observes that Berry has "developed a moral vision of man in harmony with the land."

Born in 1934 in Henry County, Kentucky, Berry grew up in a rural area of the Kentucky River near his present home at Port Royal. His respect for work and nature, his love of place, and his awareness of community were forged during these years. Graduating from the

University of Kentucky with an A.B. and an M.A. in English in 1957, he taught at Stanford University and later at New York University. But the pull of his boyhood experiences eventually drew him away from the academic and literary world of the East Coast, back to his childhood home in Kentucky. There, in 1964, he returned to the land, combining farming and writing with teaching at the University of Kentucky. Out of his experiences, out of the lessons he learns as he seeks to regenerate and heal the depleted soil of his own farm, he poses urgent questions: "What is this place? What is in it? What is its nature? How should men live in it? What must I do?"

Berry's voice is quiet and measured, the voice of a man seeking answers, a man who chooses to love, commits to heal, and is tragically aware of the knife-edge on which our planet teeters. His writing—strong in imagery, rich in description, rooted in place, and reflective—moves us back to the land, urging us to understand the universality of our relationship and responsibility to it, and our choices. "We know too, from the study of agriculture," Berry tells us in *Standing By Words,* "that the same information, tools and techniques that in one farmer's hands will ruin land, in another's will save and improve it." So he warns, holds up an example, then leaves his readers to evaluate their own roles and commitments to the survival of the planet.

Wendell Berry Resettles America: Fidelity, Education, and Culture

Steven Weiland

In "The Satisfactions of a Mad Farmer," a poem on the satisfactions of a "mad" but wise farmer, Wendell Berry praised any man "whose words lean precisely to what exists, who never stoops to persuasion." Similarly he has acknowledged in another poem, "A Standing Ground," the necessity of avoiding certain kinds of public discourse: "Better than any argument is to rise at dawn and pick dew-wet berries in a cup." In the early years of his career Berry may have been uncertain about the uses of argument, but he wrote then what he is famous for now, distinctively personal essays on a wide variety of public issues. In his many volumes of essays Berry's views on farming, race relations, regionalism, social welfare, war, education, eating, word processing, and other subjects are mingled with biographical essays and criticism of some favorite writers, including Homer, Thoreau, William Carlos Williams, Edward Abbey, and Wallace Stegner. He thinks of himself primarily as a farmer and writer whose work reflects knowledge of the history of agriculture and whose writing (in all genres) is in the tradition identified by Thoreau. In what follows, I offer a highly selective account of Berry's themes, favoring his own words that criticize and seek the reform of work on the land and in our major institutions, a project of national resettlement.

The Two Estrangements

The Unsettling of America (1977) is a representative text in which Berry explores these major themes: agriculture considered historically and in its present state, marriage and domesticity, and the uses of education. Much of what he says is in response to the question he posed in "The Silence," a poem in *Farming: A Handbook* (1970): "What must a man do to be at home in the world?" He must, he asserts in *The Unsettling of America,* find personal solutions for the crises we face: of character, agriculture, and culture. All three are the result, Berry claims, of "the abstract values of an industrial economy preying upon the native productivity of the land and its people." For we are in the midst of an "exploitative revolution" whose first victims are character and community, or their integration in culture as Berry understands it. Their neglect is a reflection of our disregard for the land and its abuse, especially by what is now known as agribusiness. Individual and social life are made and remade as expressions of their local origins: "Our culture must be our response to our place, our culture and our place are images of each other and inseparable from each other, so neither can be better than the other" (22).

One purpose—certainly a persuasive one—of *The Unsettling of America* is to make a compelling case against federal agricultural policy and, on the state level, the unhappy effects on American farming of the land grant universities and their schools of agriculture. Berry identifies our hunger for "specialization" and uncritical deference to "expertise" as the consequences of a misguided farm policy and educational philosophy, both indifferent to the interrelation of culture and agriculture.

> If we conceive of a culture as one body, which it is, we see all of its disciplines are everybody's business, and that the proper university product is therefore not the whittled down, isolated mentality of expertise, but a mind competent in all its concerns. To such a mind it would be clear that there are agricultural disciplines that have nothing to do with crop production, just as there are agricultural obligations that belong to people who are not farmers. (43)

The character and products of the whole culture must be understood to reflect their agricultural origins. As the values of farming and agriculture become defined by the specialist mentality as only economic, the results of all labor turn bitter and fruitless. With specialization comes greater social intricacy but less social structure, more organization but less satisfying terms for order.

> The community disintegrates because it loses the necessary understandings, forms, and enactments of the relations among materials and processes, principles and actions, ideals and realities, past and present, present and future, men and women, body and spirit, city and country, civilization and wilderness, growth and decay, life and death—just as the individual character loses the sense of a responsible involvement in these relations. (21)

This definition of the requirements of an active culture is the cornerstone of Berry's thought: "A culture is not a collection of relics and ornaments but a practical necessity" (43).

In the long chapter in *The Unsettling of America* on "The Body and the Earth," Berry finds that our several cultural crises are all based on the "isolation of the body from the many specialized activities which dominate everyday life and from all other living things." In his view modern work is too abstract and our relation to the land defined mainly by what can be taken from it. Our bodies, he says, are too weak and joyless: "Contempt for the body is unavoidably manifested in contempt for other bodies—the bodies of slaves, laborers, women, plants, and the earth itself. Relationships with other creatures become competitive and exploitative rather than collaborative and convivial" (105). We are now divided within ourselves and from each other and the land. These divisions are identified as first sexual and then ecological—the most important divisions because the most fundamental.

The divisions can be bridged and critical connections reestablished through "fidelity" considered as a cultural discipline. Berry's interest in this most traditional of virtues is based on his belief in its social practicality and its relation to the conservation of energy.

> At the root of culture must be the realization that uncontrolled energy is disorderly—that in nature all energies move in forms; that, therefore, in a human order energies must be *given* forms.... Fidelity can thus be seen as the necessary discipline of sexuality, the practical definition of the moral limits within which such responsibility can be conceived and enacted. (122)

Our deepest expression of fidelity is in marriage, which Berry sees as the central cultural bond on which many others depend. He blends a realistic view of its rhythms with an insistence on its symbolic force.

> What marriage offers—and what fidelity is meant to protect—is the possibility of moments when what we have chosen and what we desire are the same. Such a convergence obviously cannot be continuous. No relationship can continue very long at its highest emotional pitch. But fidelity prepares us for the return of these moments, which give us the highest joy we know. (122)

Berry's reputation may be said to have been established with the emergence of the counterculture in the 1970s. But the sexual revolution that came with that cultural transformation has always signified for him the potential for major cultural disruptions. In *What Are People For?* (1990) he dismisses it as an "industrial phenomenon" making a "pleasure machine" of the body: "Like any other industrial enterprise, industrial sexuality seeks to conquer nature by exploiting it and ignoring the consequences, by denying any connection between nature and spirit or body and soul, and by evading social responsibility" (191). Berry has carried this argument into defending his refusal to write with a computer, believing as he does that the physical dimensions of language use—voice and longhand—are indispensable to its meanings. Language is an art of the body. "Does shaping one's words with one's own hand impart character and quality to them, as does speaking them with one's own tongue to the satisfaction of one's own ear?" But the body of the artist or writer is the vehicle too of fidelity to the past: "All good human work remembers its history" (192).

Berry often asks for recognition of the continuity between his literary and agricultural interests. His monograph-length essay on "Poetry and Place" (in *Standing By Words* [1983]) derives from a sentence that has long lingered in his mind. It describes the relation between his work on his small Kentucky hill farm and his work as a poet: "This place has become the form of my work, its discipline, in the same way the sonnet has been the form and discipline of the work of other poets: if it doesn't fit it's not true" (92). Still, many supporters of Berry's views on agriculture and the natural world are perhaps less dedicated to his literary criticism. And no doubt many find his conviction on certain necessities of domesticity unappealing. From his point of view, however, they are inseparable. Private and public life, love and work (in his case the work of writing too) are bound historically and practically by the household. The failure of marriage and abuse of the land constitute for him the "two estrangements" most responsible for our sense of cultural loss.

"Fidelity" to the land is maintained as part of what is termed the "necessity of wildness," part of the natural order. This second sense of fidelity is needed because wildness is the permanent context for good farming as instinctive sexuality is the context for marriage. And therefore

> fidelity to the natural order preserves the possibility of choice, the possibility of the renewal of devotion.... One who returns home— to one's marriage and household and place in the world—desiring anew what was previously chosen, is neither the world's stranger nor its prisoner, but is at once in place and free. (180)

Berry favors a form of domestic and cultural life as traditional as it is, now, near revolutionary.

It includes too a third form of fidelity, to the forms of language. What threatens our language is our preoccupation with technology and the future. Mindful of how the term *love* is overused, abused, and sentimentalized, he proposes that it guides language as the body (literally) shapes it. "Only the action that is moved by love for the good at hand has the hope of being responsible and generous. Desire for the future produces words that cannot be stood by. But love makes language exact, because one loves only what one knows." Berry has refreshing admiration for the literary canon—Dante, Milton, Wordsworth—and he believes that the forms of literature represent fidelity to the best uses of language: particular, disciplined (by the past), reciprocal, generous. To stand by words is to be responsible for how they must be brought to express new cultural relations: "If we are to have a culture as resilient and competent in the face of necessity as it needs to be, then it must somehow involve within itself a ceremonious generosity toward the wilderness of natural force and instinct" (*Unsettling of America,* 131).

Unsettled by Education

Berry has had an unorthodox academic career, but he speaks with authority on the matter of the place of agriculture and education in the university. While other critics of higher education have focused on the fate of the literary canon or on the politics of multiculturalism, Berry has taken up a little-noticed theme: the historical evolution of the land grant ideal, a formative feature of many of the nation's largest universities. In

The Unsettling of America he is blunt about the "betrayal" that has seen the agricultural colleges (together with their university partners the agricultural experiment stations and the cooperative extension services) put the interests of agribusiness at the heart of agricultural education.

Even apart from this capitulation, the land grant universities—now understood to reflect the parochialism of their specialized agricultural units—have been a "degenerative" influence on higher education as a whole. According to the famous Morrill Act of 1862, these institutions, with funds gained from the sale of public lands given to the states by the federal government, were to offer a "liberal and practical education." Berry sees in the intended combination an unacknowledged recognition of Jeffersonian ideals of learning for participation in democracy, while Morrill himself (a Vermont senator) favored occupational utility. It has, however, been the fate of the land grant universities to ignore the first for near exclusive attention to the second. Preoccupied as they are with the "standard of practicality," they have accepted almost any subject and any form of learning on behalf of the "changing world" facing their current students and the need to constantly adapt to its economic demands. Berry risks a strongly stated binary view.

> It could be said that a liberal education has the nature of a bequest, in that it looks upon the student as the potential heir of a cultural birthright, whereas a practical education has the nature of a commodity to be exchanged for position, status, wealth, etc., *in the future*. A liberal education rests on the assumption that nature and human nature do not change very much or very fast and that one therefore needs to understand the past. The practical educators assume that human society itself is the only significant context, that change is therefore fundamental, constant and necessary, that the future will be wholly unlike the past, that the past is outmoded, irrelevant, and an encumbrance upon the future—the present being only a time for dividing past from future, for getting ready. (157)

When he modifies his position, it is not in order to accept the educational ideology of "the changing world" but to remind his colleagues in the arts and humanities of the gifts of practical education (as understood by both Jefferson and Morrill). Hence: "Without the balance of historic value, practical education gives us that most absurd of standards: 'relevance,' based upon the suppositional needs of a theoretical future. But liberal education, divorced from practicality, gives something no less absurd: the specialist professor of one or another of the liberal arts, the custodian of an inheritance he has learned much about, but nothing from" (*Unsettling of America*, 158).

Having dissociated himself from the current land grant establishment, Berry is no less kind to the advocates of liberal learning who have also, in their habits of abstraction and social ambitions, subverted the cultural meaning of higher education. He acknowledges how much information and knowledge the university produces but asserts that its professional instincts of "self-esteem" misdirect the benefits: "We do not work where we live, and if we are to hold up our heads in the presence of teachers and classmates, we must not live where we come from" (*Unsettling of America*, 160).

When Berry returns to the theme of higher education in *What Are People For?* it is to register the continuing conversion of the university into an international "economic resource" that has the effect—despite its sloganizing on behalf of academic "service"—of continuing to undermine traditional rural life and the complex network of cultural values it sustains. Berry accepts the biological and social roles of competition within limits. The land grant universities, having adopted economic determinism and the unexamined ethics of competition, have ignored how change might be balanced with "constancy." Higher education requires fidelity too, to ideas and to texts, and to resources for continuity in liberal learning. Many professors in the liberal arts understand themselves to have been struggling to sustain such ideals within the land grant university. But they have had no impact on the "progress" of agricultural education, and for this they have earned invective that is unusual for Berry (apart from what he reserves for his "industrial" targets): "The careerist professor . . . blunts his critical intelligence and blurs his language so as to exist 'harmoniously' within [his institution]—and so serves his school with an emasculated and fragmentary intelligence" (120). Berry urges that the academic disciplines be freed from their competitive struggles and reassembled to serve "place" and other values.

Berry's attack on the academic vocations reflects his own fastidious standards for work. But it reveals too a lack of generosity for practices that, while they may be timid in relation to economic and political interests, are also necessary to the preservation of culture. That role is part of a historical formation that Berry both reflects and resists. He has joined others (e.g., Russell Jacoby in *The Last Intellectuals* [1987]) in noting the shrinking of intellectual roles since the 1950s, but he is alone among our cultural critics in seeing how education for agriculture, and what it reflects about attitudes toward the land and its real value(s), has prompted a larger educational default, this one signifying the inability of our professors to secure a better integrated and more genuinely critical vocation. Where there should be academic "fidelity," especially to one's educational place, there is only misguided freedom: "One's career is a vehicle, not a dwelling; one is concerned less for where it is than for where it will go" (*Unsettling of America*, 148).

What Is Culture?

Academic life is a place for language, but without understanding of its consequences. Those who appear to "stand by words" do so within self-imposed constraints. They might ask (as Berry proposes in *Standing By Words* in an unusually schematic presentation of the "system of systems"), "How appropriate is the tool to the work, the work to the need, the need to other needs and the needs of others, and to the health of the household or community of all creatures?" The resettlement of America depends on a productive conflict between modern social (domestic), agricultural, and educational practices and traditional but now "marginal" alternatives.

> As an orthodoxy loses its standards, it becomes unable to measure itself by what it ought to be, it comes to be measured by what it is not. The margins begin to close in on it, to break down the confidence that supports it, to set up standards clarified by a broadened sense of purpose and necessity, and so to demonstrate better possibilities. (51)

Berry frequently cites the Amish farmers as examples of the critical margin since they suggest a standard against which today's culture and agriculture can be judged. "I do not see," Berry says, "how a stable, abundant, long-term agriculture can be built up and maintained by any standard less comprehensive than that of the perfect health of individual human bodies, of the community, and of the community's sources and supports in the natural world." This is a high standard demanding that culture be the perfect expression of a balanced social and economic order.

It should be built on a diversified system of farming supported by suitable forms of teaching and learning widely distributed throughout the society. Culture is at the same time the source and the product of agriculture. Accordingly, it is always local, though it reaches across place and time: "A healthy culture holds preserving knowledge *in place* for a long time. That is, the essential wisdom accumulates in the community much as fertility builds in the soil" (*Standing By Words,* 73).

Culture is a term Berry often uses to identify the gathering of traditions and human qualities he admires. He is, of course, only one of many who now make "culture" the subject of critical comment and study. The number of books and essays in the humanities and social sciences with *culture* in their title is, however, more a suggestion of the ambitions of the authors than proof of agreement about what the term actually means. The explanation of culture is a field dominated by anthropologists who, predictably, disagree among themselves about what exactly their field explains. Clifford Geertz, for example, reminds us that even Clyde Kluckholn, in his influential *Mirror for Man* (1949), manages to define culture in almost a dozen different ways. Geertz himself proposes in *The Interpetation of Cultures* (1973), now of course more widely known than Kluckholn's famous book, that "man is an animal suspended in webs of significance he himself has created and culture is these webs" (5). A diffuse but important form of inquiry called interpretive social science can be said to have emerged from this position, as has, also, the academic practice called cultural studies. Both represent recognition that "culture" as a category for inquiry is larger than the resources of any single academic discipline and that new scholarly configurations are needed to master its complexities. Berry shows no interest in the academic debate over the meanings of culture and the best ways to study it. His understanding is based essentially on his experience as an observer and participant.

Berry sometimes appears to relish his academic isolation, but it is important too to see how his work is continuous with forms of critical inquiry. British social and literary critic Raymond Williams specialized (in a form even Berry might admire) in studies of culture that are close in spirit to the Kentuckian's while at the same time very deliberately theoretical. His work stands behind "cultural studies." Williams's historical review of the definitions of culture suggests the durability of Berry's approach. *Culture and Society* (1958) and *The Long Revolution* (1961) are Williams's efforts to describe the important processes of change over the past two centuries: the democratic, industrial, and cultural revolutions. Each, of course, is a complex social process whose meaning can be understood through attention to the historical development of the terms of critical explanation. Hence the meaning of culture is actually a record of the changes in the use of the word. Prior to the nineteenth century it had meant, primarily, "the tending of natural growth," and then the processes of human training. The Victorians introduced the idea of "culture" as such, as opposed to the culture of something: "It came to mean, first, a 'general state or habit of mind,' having close relations with the idea of human perfection. Second, it came to mean 'the general state of intellectual development in society as a whole.' Third, it came to mean 'a whole way of life, material, intellectual, and spiritual'"(xiv). Williams himself actually came from a rural Welsh background surprisingly similar to Berry's, as he demonstrates in his novel *Border Country*. He values the traditions of agriculture and rural living as he does those of learning and the city.

Within his historical analysis Williams finds the violation of important political principles as critical as the abandonment of certain agricultural habits. He identifies "solidarity" as the key to cultural vitality, as the real basis of society. Individuals, Williams argues, verify themselves primarily in their communities and require a participatory culture which must be guaranteed by the fundamental principle of "equality of being." When Williams argues for a common culture, he is interested less in the individual or domestic expression of its virtues and products than in the collective results. All modern cultures are complex, and each needs a principle of order: "At root, the feeling of solidarity is the only conceivable element of stabilization in so difficult an organization" (*Culture and Society*). The influential American philosopher Richard Rorty has adopted a similar argument as part of his revival of Deweyan pragmatism. Any hope, therefore, for a common culture depends on the actions of a self-conscious community reinforced by suitable political structures, or the Jeffersonian model for living and learning. Culture is a source and product of any such arrangement; its analysis is the work of anyone interested in the shape and direction not only of the arts and learning but of politics and society.

In *The Long Revolution* Williams supplements his elaborate etymologies with an authoritative review of the purposes and content of the categories of cultural analysis. He names three in a synthesis of the interests of many of those whose use of culture is not always clear. The term will probably never have a precise and widely agreed upon definition, and that is as it should be; but Williams, in this lengthy quotation, helps us to understand the relation of intentions to materials.

> There is, first, the "ideal" [style of cultural analysis] in which culture is a state or process of human perfection, in terms of certain absolute or universal values. The analysis of culture, if such a definition is accepted, is essentially the discovery and description, in lives and works, of those values which can be seen to compose a timeless order, or to have permanent reference to the universal human condition. Then, second, there is the "documentary," in which culture is the body of intellectual and imaginative work, in which, in a detailed way, human thought and experience are variously recorded. The analysis of culture, from such a definition, is the activity of criticism, by which the nature of the thought and experience, the details of the language, form, and convention in which these are active, are described and valued.... Finally, third, there is the "social" definition of culture, in which culture is a description of a particular way of life, which expresses certain meanings and values not only in art and learning, but also in institutions and ordinary behavior. The analysis of culture, from such a definition, is the classification of the meanings and values implicit and explicit in a particular way of life, a particular culture. (41)

Williams's summary of the styles of cultural criticism provides an excellent framework for understanding Berry's distinctive American contribution. He makes forceful judgments in all three categories. His is an unusually comprehensive display of cultural criticism, but without the apparatus of professional theorists. Berry has no general theory of cultural analysis, only the particular instruments of its practice: a commitment to high ideals of human behavior; analytical interest in the arts and learning, including attention to literature and history; and daily attention to the traditions and routines of everyday life, especially of agricultural and domestic life.

Conclusion: Making the Old New

Though he may admire the personal or domestic values of "fidelity" rather than the political values of solidarity, Berry also believes in the need for a common culture. He stresses the development in individuals of the disciplines necessary to healthy family and community life: fidelity in one's relations with other people, with the land, and in the organization of education and the uses of language. These reveal in a productive culture the cyclical nature of human life: "It is only in the processes of the natural world and in analogous and related processes of human culture, that the new may grow usefully old and the old made new"(*A Continuous Harmony*, 150). For these reasons culture is never progressive, but neither is it simply the cumulative display of art and learning. It is the network of disciplines, as Berry defines them, applied to the facts of everyday life and the possibilities for the future.

"The great moral labor of any age," Berry says, "is probably not in the conflict of opposing principles, but in the tension between a living community and those principles that are the distillation of its experience" (*A Continuous Harmony*, 153). It is the purpose of his books to mediate those tensions as they explain them. His methods of cultural analysis are decidedly less elaborate than those of either Raymond Williams or contemporary anthropologists, but his findings are equally pointed and, because of his prescriptive posture, more practical. Solidarity, finally, suggests the primacy of social answers to questions of value facing individuals. Fidelity is preeminently a trait of healthy individuals and citizens. Berry stresses (re)settlement as the goal of private and family life and the source, therefore, of a society firmly rooted in its own homemade culture.

One of the pleasures of Berry's books is the visibility they give to his friends and neighbors who reflect in their everyday behavior his discursive preoccupations.

> I was fortunate, late in his life, to know Henry Besuden of Clark County, Kentucky, the premier Southdown sheep breeder and one of the great farmers of his time. He told me once that his first morning duty in the spring and early summer was to saddle his horse and ride across his pastures to see the condition of the grass when it was freshest from the moisture and coolness of the night. What he wanted to see in his pastures at that time of year, when his spring lambs would be fattening, was what he called "bloom"— by which he meant not flowers, but a certain visible delectability. He recognized it, of course, by his delight in it. He was one of the best of the traditional livestockmen—the husbander or husband of his animals. As such he was not interested in "statistical indi-

cators" of his flock's "productivity." He wanted his sheep to be pleased. If they were pleased with their pasture, they would eat eagerly, drink well, rest, and grow. He knew their pleasure by his own. (*What Are People For?* 140–41)

Observed against the work of agribusiness and his academic habits too, Besuden's knowledge and skills have the virtues of their apparent faults; he neither competes nor (excessively) contemplates. Thus his achievements stand for the ways in which America might be resettled according to fresh (if old) values. Writing about the responsibilities of poets (in *What are People For?*) Berry asserts that

> Professional standards, the standards of ambition and selfishness, are always sliding downward toward expense, ostentation, and mediocrity. They tend to always narrow the ground of judgment. But amateur standards, the standards of love, are always straining upward toward the humble and the best. They enlarge the ground of judgement. (90)

Henry Besuden's "pleasure" is his contribution to the local culture. But so too does Berry find his examples in the pleasures of (even "professional") literary tradition, especially when it expresses other indispensable forms of culture. Here, in a sentence Berry endorses, is Emerson speaking to Berry's pleasure and, across history, about the latter's unique late twentieth-century vocation with its mingled solidarity and fidelity: "I grasp the hands of those next me, and take my place in the ring to suffer and to work, taught by an instinct, that so shall the dumb abyss be vocal with speech" (*What Are People For?* 85).

Wendell Berry's Husband to the World: A Place on Earth

Jack Hicks

A farmer and professor, Wendell Berry has also had a prolific literary career since his first book, *Nathan Coulter* (1960), and he works in all forms—poetry, fiction, essays, and drama. From his artistic beginnings, he has shown an abiding interest in his central Kentucky homelands. And like his fellow novelist of the land, Ernest J. Gaines, who sets his tales in the dust and bayous of rural Louisiana, Berry also tills a single native soil. His constant terrain has been the Upper Appalachian South, in the locale of Port William, Kentucky (a poetic imagining of his own Port Royal), spread across the rolling hills and cuts that drain into the Kentucky River. Berry's fictive families, mainly the distantly related Coulter and Feltner clans, are subsistence tobacco farmers and dwell in all three of his novels and much of his other work. Their stories are set mostly in the 1940s and 1950s, but they range as far back as the antebellum Simon Feltner (1784-1858) and span at least seven generations—actively and historically—in the novels.

"The earth is the genius of our life," Berry writes near the end of *A Place on Earth*, and from that earth and a sense of man's place in it, he has developed a moral vision of man in harmony with the land, a conservative, Jeffersonian agrarian ideal rare and attractive to our times.[1] He has had help and pays frequent witness to his psychic kinsmen, to Jefferson and Thoreau and, more recently, to the Southern Agrarians of *I'll Take My Stand*, to William Carlos Williams and Gary Snyder.[2] Like that of his former Stanford classmates Gaines and Ken Kesey, Berry's work is centrally concerned with "the genius" of modern man in a specific geographical place, and his work speaks to those who yearn for a healing vision of the mingled lives of humans and nature. He has gone back to the old ways, and the richness of his publication (*Esquire, Harper's, New York Times, Hudson Review* on one hand; on the other, *Mother Earth News, Organic Gerdening and Farming,* Sierra Club Press) testifies to the broad appeal of his message.

The model of Berry's own life, recounted in the departures and returns to his family and calling and place in *The Long-Legged House* and in the recent poetry—especially in his restoration of the family "Lanes Landing Farm" in 1965—has nourished and been nourished by an extraordinary rich metaphor: man as husband, in the oldest senses of the word, having committed himself in multiple marriages to wife, family, farm, community, and finally to the cycle of great nature itself. This is the central stream of Wendell Berry's writing, his "country of marriage" (the title of a recent book of poems), the controlling pattern of his imagination, and it travels richly through the images, languages, and tales of his work, just as the Kentucky River winds through the loamy tobacco lands that so possess his imagination.

Whether we speak of his polemics against the twin rapacities of strip-mining and agribusiness (Berry is among the clearest contemporary ecological voices), his practical essays in *Organic Farm and Gardening,* his lyric celebrations of the feel of rain and wood and friendship, or his brooding tales of seven Port William generations, the informing vision is the same, a complex and coherent sense of man's need for a proper place on earth. His assumptions are unstated, at times in conflict, but in essence his view of man is as a distinctly flawed being fallen from natural wholeness. A ruined forest kingdom lies faintly in the background of Berry's work, idyllic and Edenic, a prelapsarian, preagrarian world of unspoiled nature. A version of the destruction of that primal world is offered in a recent narrative poem, "The Kentucky River: July, 1773."[3] This is the earliest historical moment in his Kentucky valley saga and a reenactment of an ancient tragedy. Berry often links the voracious westering impulse in American history with the primal violation of nature, and here we see the first white explorers from Virginia, incredulous at the serene vitality of sacred Indian lands near Big Bone Lick. While Berry does not stress the Christian element—indeed his work shows little sympathy with organized religion—his settlers are as near the garden as mortal man can ever return, held spellbound, "for that upswelling / and abounding, unbidden by any / man, was powerful, bright, / and brief for men like these, / as a holy vision" (9).

Young Sam Adams finds himself, entranced, in an unfrightened herd of grazing buffalo, a young player in an old scene, and fires his musket in their midst. His name suggests his role, and to Berry's mind, he is at once the first Adam falling from unity, the westering white man wasting nature in his path, and a boy trumpeting the arrival of male sexuality. They are links in a chain of motifs related regularly in the work. The entire party is nearly trampled, and Berry painfully searches for motives, concluding: "He was faced with an amplitude / so far beyond his need / he could not imagine it, / and he could not let it be. / He shot" (9). Thus man, particularly the historical American white man, lives fragmented since his fall from harmony with nature, divided, condemned to "obscure or corrupt our understanding of any one of the basic unities."[4] Berry's ideal husband is earthly man in his most noble state, doomed to separate consciousness, but in that single mind making a pact with the world, taking the vows of marriage, assuming the healing role of husband to wife, family, and land. Farm, community, family—these are earthly compromises, the tropes in flesh and word and wood of mortal man, his ritual gestures to re-create whatever harmony he can. The farm itself is not a "natural" shape, but "an opening in a wilderness," a man-attended order, often an attempt by Berry's farmers to heal the geographical and historical scars of their wrongheaded ancestors.

The husband's literal and metaphoric role, as the language and substance of what Berry terms "the metaphor of atonement" suggest, is one of healing old wounds, of atoning for past violations, by the reawakening in human consciousness of the sense of nature's "interlocking systems" (*A Continuous Harmony*, 157), of the possibilities that human lives might share and thrive in the old organic dream ("at-one-ment," as a later gloss suggests, 159). As we see later, this is an ideal, and ideal husbands, as Mat Feltner of *A Place on Earth* seems, are as rare as any ideal. Much of the tragedy and pathos in Berry's work originates in the failure—either willed or fated, conscious or unaware—of men to perceive a natural order or conduct their lives within it.

Pervasive though it is, the "metaphor of atonement" is most directly shaped in the essays of *A Continuous Harmony*. Here Berry writes of the sacred bonds between man and land, of the marriage of the husband to his literal and mystic wife. In "Think Little," he invokes Black Elk, holy seer of the Oglala Sioux, and his visions of the interconnectedness of all life: "I saw that the sacred hoop of my people was one of many hoops that made a circle, wide as daylight and as starlight, and in the circle grew the mighty flowering tree to shelter all the children of one mother and father. And I saw that it was holy."[5] "Discipline and Hope," his most ambitious essay, calls several witnesses: Inca historian John Collier notes the tribal unit, the *ayllu,* is based in "not merely its people and not merely its land, but the people and the land wedded through a mystical bond"; and Rhodesian Tangwera Chief Rekayi calmly refuses to cede ancestral land to whites, explaining: "I am married to this land. I was put here by God" (102).

Sioux, Inca, Tangwera—these are tribal cultures, "primitive" images of noble pasts, Berry chases the thread closer, to Thomas Jefferson, whom he quotes as describing farmers as "tied to their country, wedded to its liberty and interests, by the most lasting bonds" (104). It is finally in "Discipline and Hope," against this historical chorus, that he most clearly shapes his moral and literary vision.

> Living in our speech, though no longer in our consciousness, is an ancient system of analogies that clarifies a series of mutually defining and sustaining unities: of farmer and field, of husband and wife, of the world and God.... A man planting a crop is like a man making love to his wife and vice versa: he is like a plant in the field waiting for rain.... All the essential relationships are comprehended in this metaphor. A farmer's relationship to his land is the basic and central connection of the relation of humanity to the creation; the agricultural relation *stands for* the larger relation. Similarly, marriage is the basic and central community tie; it begins and stands for the relation we have to the family and to the larger circles of human association. And these relationships to the creation and to the human community are in turn basic to, and may stand for, our relationship to God—or to the sustaining mysteries and powers of the creation.... If the metaphor of atonement is alive in a man's consciousness, he will see that he should love and care for his land as his wife, that his relation to his place in the world is as solemn and demanding, and as blessed, as his marriage; and he will see that he should respect his marriage as he respects the mysteries and transcendent powers—that is, as a sacrament. Or—to move in the opposite direction through the changes of the metaphor—in order to care properly for his land he will see that he must emulate the creator; to learn to use and preserve the open fields... he must

study and follow natural process; he must understand the *husbanding* that, in nature, always accompanies providing. (159–61)

My interest here is in the continuity of Berry's vision, and particularly in how the image of the exemplary husband—in his ideal and lesser aspects—is refracted in his novels, most especially in the fullest and most satisfying fiction to date, *A Place on Earth.*

The world of Port William is male-dominated, and the first depiction of it is in *Nathan Coulter,* a spare bildungsroman made up of the protagonist's episodic recollections of his formative years, roughly to age fourteen. The lessons of manhood and the instructions of husbandry are hard-learned and seldom gentle. In the figure of young Nathan, Berry depicts an *apprentice-husband,* one who will discover the many stern vocations of farm marriage. Filtered as it is, back through a youthful first-person consciousness, the novel shows little of the rich verdure of history that characterizes Wendell Berry's best work, few of the intertwinings of characters—their lives and pasts—that suggest the ripe weight of past on present.

Young Nathan's stark recollections are truly of a ruined kingdom, for what he remembers are not so much the "sacred hoops" of man and wife, family, or the worlds of the fields and the woods. His discoveries are not of what is but of what is lacking. From his earliest sense of separate consciousness ("I'm Nathan Coulter. It seemed strange."),[6] his memories are suffused with awkwardness and alienation and finally multiple loss—of innocence, of parents and siblings, of community. The emphasis is on the rending of the organic fabric, and this apprentice sees many failures of the dream of harmony, a long sequence of disrupted relationships between husbands and the natural world.

As in much of Berry's work, the major thematic interest is in death, in Nathan's gradual awareness of it, in how one deals with it or does not, and death is omnipresent: in the natural deaths of fish and game that he and his mentor, Uncle Burley, harvest in the dramatic background; in the death of his own pained childhood; in the more metaphoric deaths of the body of the family and his father's spirit.

Nathan cannot accept the role of husband to this damaged world (although he has returned five years later in *A Place on Earth*), and as he leaves near the end of *Nathan Coulter,* he takes a last look backward to see a mirroring of several deaths: "I could see them all through the window, sitting with Daddy by Grandpa's coffin, keeping their separate silences, their faces half shadow in the dim light" (203).

Ninety-two-year-old Jack Beechum, central figure of *The Memory of Old Jack,* is the *master* to the youth's apprenticeship, a declining husband of the old ways, one whose death raises "the possibility that men of his kind are a race doomed to extinction."[7] The "memory" of the title is doubly significant. First, the novel is indeed a memory, an elegy and a requiem for the oaklike old man whose final earthly day in September 1952 is the fictional present. For the men working in the tobacco shed three months later—the Feltners and Coulters and Penns and Catletts— he has indeed been "a monument . . . a public statue" (3-4), an emblem of the husband and his legacy. The novel closes on their agreement that death can end a life but cannot cancel it, that the remembered substance of Old Jack Beechum will be as rich as his physical presence, "that the like of him will not soon live again in this world, and they will not forget him" (223).

It is also "memory" in a second sense, for Beechum's last day is a series of journeys back into memories of the past, interlaced with those of friends and relatives in Port William. The main substance of the novel is precisely this mixture of reminiscence and reflection, of the marbled history of the aged husband, his work and town and land and tragic marriage. The growth of nascent consciousness kindles our interest in the future of the protagonist; the lying down of old age cants our attentions back to an earlier time. Youth and old age, *Nathan Coulter* and *The Memory of Old Jack* are endpieces, and between the two stands Wendell Berry's most developed image of husbandry, *A Place on Earth.*

The emphasis in *A Place on Earth* is on the mature husband and, as the title and recurrent language of the novel suggest, on his *place* within the organic cycles of nature. The pattern here, both implicit and explicit, is the seasonal cycle of animal and vegetal life—of sowing, germination, fruition, death, decay—and Berry develops the life of his central character, Mat Feltner, and the various marriages and families and farms of Port William, and of the larger social worlds beyond, in terms of how they do or do not partake of these natural cycles.

Feltner is the ideal husband to the world, a striving upward in the flesh, back toward unity with the natural world. He is defined, dramatically, by his struggle with the meaning of his son Virgil's death, and—more statically—by exemplary images and tales of lesser men around him. They take many shapes. They are men like Simon Crop, who because of their fate and weakness make a more tenuous pact with woman and farm. Or they are Berry's pathetic or tragic cases, the flawed and failed husbands—like Jarrett Coulter or Jack Beechum—who have failed to find the continuities they sought. Or they are bachelors, like Jayber Crow, Burley Coulter, and Ernest Finley, who have refused or been refused the varied healing roles of the husband.

From the time the novel opens, the setting—the interrelations of weather and season and place—is not merely a backdrop but an active presence. "The seedbins are empty," we start, and it is here that the arcs of seasonal growth and fictional development start: in stasis and darkness ("time as a succession of nights"), as a drear late winter rain, "the very presence and noise of emptiness," drums on the tin roof overhead (3). It is early March 1945, and four men (Mat Feltner, Frank Lathrop, Jack Beechum, Burley Coulter) play a desultory game of rummy, "expectant of sound... anticipating an arrival," waiting for the stalled coming of the planting season. "They're waiting," we note, "for the war to be over, for whatever resumption of continuities and certainties will take place at the end of it" (13). By the close of the novel in late autumn 1945, the seasons will have turned again. The attended or unattended lives of forest and croplands, of livestock and orchards, of men and women and their families, will also have turned within that cycle, some with it for good, some against it for tragic ends. And the world of Port William and the life of the husband will return to the rain and darkness from which they issued, with a renewed sense of rest in winter.

The opening anxious gloom is appropriate psychic weather for Mat and Margaret Feltner, for they have had no word from their son, at war in Europe. The letter arrives, confirming Virgil "missing in action," and much of the body of the novel is taken up with the tensions and meanings of his loss, both to the Feltners and to his pregnant wife, Hannah, who lives with them. Though Virgil's death seems unnatural set against the coming spring, it is not unusual at this stage in the life of Port Williams for a recurring motif is the loss of the young to battle and catastrophe. Tom Coulter and Virgil die in Europe. Nathan Coulter leaves early in the action, and Jasper Lathrop and Billy Gibbs also serve. Young Annie Crop will be swept away later in a violent spring flash flood. It is as if the future of the town, its young life, is being amputated. Their losses are wounds to the social body, and Berry continues that image in a rich pattern of detail throughout the novel. The languages of healing and scarring prevail; farmers wound and scar their land and wives in ignorance; whole lives (like Ernest Finley's) are weak healings over mortal injury; Virgil's loss is a trauma to the social body and psychic life of the husband. Set against the coming greenness, all such destructions are difficult to reconcile, seem almost moral violations to the natural order.[8]

Mat tries to distract himself, but the loss of his only son threatens his entire life. He has lost "a sense of continuity," we are told, and he reaches out for "life, more purely than he ever conceived it before—his son's life and his own—restored, healed, made whole" (331). Virgil's death seems to cancel his own being, and even the trees and buildings are totemic, ghostly "monuments to a failed past" (25).

As Mat Feltner fingers the ragged wound of Virgil's death, his mind loops back through his own past. Again Berry works in cycles, here to most basically suggest the historical roots of the husband's past and to depict his many intertwining marriages. The richness of *A Place on Earth* is its accumulation of such retrogressions into individual pasts (just as the soil lives as a present corpus of past leaves and bodies), and in Mat's chapter, Berry underscores the most dominant concerns of the husband's life: work, history, marriage itself.

Mat remembers himself as a boy in the country, romping with friends "wild as foxes," and, as we are reminded in a long historical sketch, Port William was also young and unshaped, "possessing a certain wildness about it" (147). For a time, he lives "free of his life," but he comes to serve a kind of "apprentice manhood," first in daredevil play in the river bottoms, and later, in reality, at age fourteen, when his father introduces him to the sweat of the field: "I want you to learn what work is, how it's done, and how a man makes himself able to do it. If you don't learn that as a boy when you've got energy enough . . . you'll never learn it as a man" (156).

So Mat comes to tobacco farming as a youth; and, through the punishing work—from the tilth of the black soil breaking open under the plow, to the cutting and sorting of the winy burley into hands and sticks, to the final disking of the field—he becomes a man. Work is important in Berry's world, testifying to a man's relationship to nature; ideally, as exhausting as it is, the husband's labor is an entering into the rhythms and harmonies of natural growth. Work is a song attended by music in these novels, a poetic celebration of the escape from the solitary self to temporary wholeness, an action by which the husband nourishes and is nourished by his wife, the land. So Mat learns to value the sweat shed by men stooping in rows and to ease his labor in work chants ("Hundred dollars waiting on a dime. / Show it to me boys. / Make me know it." [156]).

Mat is significantly most fulfilled in the literal work of plant and animal husbandry. He takes his deepest pleasures among livestock and orchards, and we see him often in barns, lower pens, and fruit fields. Birthing animals, his life takes shape before him. "He hungers for the births and lives of his animals," Berry writes; "he's more at peace with himself than he is at any other time" (115). Delivering lambs, he experiences a Brueghelesque ecstasy in "the life of things, standing up in a lamb or a plant, his vision and his justification and his blessing" (117). The prose carries the same lyric intensity in depicting him among the fruit trees. Here again, he is most alive in husbandry, on a pruning ladder surrounded by "delicately-shaped shoots of last year's growth. Loving the color and shape and feel, the whiplike life of them." And Berry concludes, "His labor with these trees has always been one of the finest and happiest of his labors" (266).

So work is a kind of kinetic prayer, a witnessing and affirming of man's active place in the natural world, and it is also a means by which history and character are revealed in *A Place on Earth*. Mat Feltner is the moral norm and approaches the ideal husband (as steward) in his ability to work within nature, and within his own farming group—Burley Coulter, Joe Banion, Big Ellis, and neighbor Elton Penn, Old Jack's "adopted" heir—share, in different styles, his balance. The labor of man is a main theme throughout the novel, and other characters and families are tested—and often found wanting—by their relationships to work. There are the slothful, like the derelict bootlegger Whacker Spradlin, for whom narrative scorn undercuts sympathy. A career drunk, he is the last fizzling of Hoss Spradlin, a hard but misguided worker. His father devours the world, would "replace a broken window with a sheet of tin, or drive a tack with an axe." Finally, his fate "was to be a user of land and work animals and tools that other men had worn out" (28). Disordered and numbly refusing all work (his customers help themselves), Whacker represents a further decline, the swing of the pendulum from his father's manic energy. Roger Merchant, Feltner's indolent cousin, is also a severed thread, a lazy caricature of the "gentleman farmer." Like Hoss Spradlin, his father, he "lived on the land like a blight on it," as if "an angel had appeared to him, saying: "There it is. Use it up. Get all you can out of it" (178). Here, too, the son is an image of familial and personal decline.

The Spradlin and Merchant clans are dying, and the scions in each case are only sons, shriveled images of historical deterioration, characterized mainly in their refusal or inability to do fruitful work. There is also the opposite imbalance, seen most vividly in Jarrett Coulter. Mat's contemporary and Nathan's father, he is an instance of increasing isolation and moral disorder. Jarrett Coulter is a stern and loveless patriarch in *Nathan Coulter,* presiding over a household that dies under his touch. His wife and father pass, and he drives his sons from his house and is left alone, finally, in a mutated relationship to his land and labor. He is one of Berry's ruined husbands, seized in a cancerous, abstract obsession with his land. A generation older in *A Place on Earth,* Jarrett Coulter's decline is more pronounced, and he again represents the masculine impulse to shape and order gone wild. His life is driven by a grim energy to harness the natural world, and his work is a kind of race, as "he bears down on a crop as a runner bears down on a tape." His dream is an exaggerated reverie of "perfect order and perfect weather," and to this unnatural end, his labor is relentless, has "created failures of friendships between him and his sons, and him and his neighbors and has carried him past them—into silence that he has made his calling and his answer" (528).

Part I, chapter 3, devoted to Mat's past, depicts a second concern of the husband—history—and again the element is active in many lives in the novel. If nature reels in continuing cycles, so too does the present life of a man or a family or a town grow from the past. History is a kind of root stock in *A Place on Earth,* and the many pasts of people and places give branch to and enrich the narrative present of the novel. Summoned directly in the strong narrative voice, or as family apocrypha over ham and biscuits at the supper table, or as town gossip in Jayber Crow's barber shop or Frank Lathrop's store, the multiple histories accumulate and serve to gloss the positive instance of Mat Feltner's life.

From childhood, through adolescence, to adulthood, Mat's own personal and family history has been a preparation for his life as a mature husband. What emerges from the pattern of his life—Berry emphasizes this—is that he finally *elects* his role, takes vows of marriage to his wife and place and people. Further, he was historically prepared for it, perhaps even *fated* for it, so that he does indeed finally "choose what he was destined to choose" (175).

And here again, the tendrils of history and fate extend through many lives. The narrative line is deflected regularly from the season of 1945, taking us into the histories of the major male characters—Jarrett and Burley Coulter, Ernest Finley, Jayber Crow, Gideon Crop. If Mat's life is a vital harmony, a balance of past and present, fate and free choice, personal duty and family obligation, there is also an attendant strain of imbalance, of historical misfortune and tragedy. Fate can write one's doom, or one can be driven by a false sense of historical imperative. Each case is symptomatic, like bad work, of unnaturalness, disorder. One of the earliest of the Feltner clan, former slavetrader and Confederate "officer" Jefferson Feltner is a case of historical myopia, another instance of the masculine need to order run unchecked. After the war, he is obsessed with "honor" and "loss of past," becomes a zealot who fights constantly to defend a misvision of person and family. His life is a martial dance, an extended, ruthless vignette of the dangers of historical illusion, and he is "the servant and instrument, and finally, the victim of his history" (152). More pitifully, Roger Merchant languishes in a besotted fantasy, "an uncritical devotion to what he called his family tradition" (179). He reaches a dead end, bachelor son of a failed and ravenous husband, alone and unwed to woman, soil, or people, a man who "has built nothing, added nothing, repaired nothing" (182). And his illusions of his scruffy father "as a cultivated and enlightened gentleman farmer" and his own "latter years of a little light fawming" are the solipsisms of the bottle, as passive a historical fantasy as Jefferson Feltner's were active, but no less destructive.

The long and tragic story of Ernest Finley's life and suicide is perhaps the most moving tale in *A Place on Earth* and drives home the lesson that history and fate play active roles in human life. A local carpenter, Finley is a meticulous craftsman, a quiet, respected friend and worker. While his suicide first seems shocking and aberrant, reflection suggests that it has been well prepared, and his self-destructive act seems finally almost unavoidable. Also one of Wendell Berry's bachelors, Finley returns to Port William after World War I, apparently to pick up his life in the community. But his ties have been irremediably severed. His parents dead, his family house and lands dispersed, Ernest has also been war-crippled; he is shown to be a man historically and personally wounded, fated to a tragic end. He is a physical and psychic cripple, described recurringly in the language of scars and scarring. His body is a metaphysical damage report: "Fragmented bones and tendons were spliced back together and packaged in scar tissue" (47). His nickname, "Shamble," is "placed on his life like a scar, the healing of a wound and its betrayal" (51). Even his apparent virtues, serenity and patience, are finally delusive surfaces, keloids like his shop, "a walling in of his desire" (52). His life is another dead end.

Following the flash flood that devastates the Crop household, sweeping Annie away (and Gideon in a crazed search for her), Ernest works neighborly to repair damaged outbuildings. Berry lingers over his shy, gradual attachment to Ida, his unwitting attempt to deny his history and fate. Gideon writes that he is returning, and the strange house of Ernest's need collapses around him, "as if sunk into the blinding whiteness of Gideon's letter" (462). His aberrant "marriage" to Ida has been historically and socially unnatural, and his suicide—at first shocking—seems finally a kind of healing. Death is a part of the natural process, the end and the beginning—this is to be Mat's lesson—and Ernest's end is a grim picture of the death of a crippled line, the healing return of blood to the natural earth. Mat Feltner finds him in a workshop corner, dead on his knees, both wrists deeply and carefully cut, his blood drained through a newly sawed hole in the floorboards.

Like Ernest, whose history seems to refuse him the easings of husbandry, even life itself, many of Berry's men are widowed or unwed. This is no paradise of bachelors: bitter Jarrett Coulter and the old chief Jack Beechum are widowers; Roger Merchant, Burley Coulter, Jayber Crow, and Ernest Finley, for all their differences, have refused or been refused the vows of husband and father. From Mat's pained broodings, he is eased as he turns to this third consideration, marriage. A man's life is a statement, and marriage and its human condition are ideally a statement in flesh of the multiple vows of the husband. Like the lives of man and town, his married life has grown from childhood to maturity. He and Margaret Feltner are childhood friends, and their long courtship is a complex process, often painful, of his learning to accept her as wife and partner. His return to Port William after college and travel is a multiple acceptance, of "his life before him, his marriage, his place, his work" (73).

Throughout the novel, the emphasis on marriage, on the joining of male and female, is decidedly on the faces of friendship, partnership, parenting. Berry draws few scenes of sexual love or passion; indeed sexuality—especially male sexuality—often threatens marital harmony. Mat's own incipient adolescent sexuality, for example, threatens his relationship with Margaret until he masters it, and "their relationship slowly healed" (164). And the protagonist's coltish discovery of male energy in *Nathan Coulter* is devastating: his first sexual experience with Mrs. Mandy

Loyd is a re-enactment of man's primal violation of natural order, ruining the Loyd marriage, disrupting his family, nearly costing him his life.

Berry's farm marriages are more characterized by quiet respect, support, and endurance, than by passion, intensity, or personal encounter. Marriage is distinctly man-made, "a practical circumstance," he writes in "Discipline and Hope," as he emphasizes the homespun aspect; "it must make a household, it must make a place for itself in the world and in the community" (103). When his wives are depicted as sexual beings, it is most often as a fruitful presence, as an embodiment of the feminine principle, the less-treated but fully complementary partner to the will to husband. The pregnant Hannah, for instance, continues life just as the death of her young husband threatens that continuity. Despondent in his loss, Mat finds solace in her presence. Big with child, she is shown among livestock and flowering orchards, healing the break, carrying and nourishing life just as she bears a spray of peach boughs, "the graceful curving and slendering of them weighted and knobbed with buds" (267).

Like work, marriage is a rising up, ideally a merging of the solitary selves, an act of healing and partial reconciliation with nature. Though the Feltner marriage is a model, many others are less successful. Apparently doomed to tenant farm Roger Merchant's land and never work their own, Ida and Gideon Crop live with fewer hopes in a weaker bond. Driven by a sense of futility, Gideon vents his despair in drinking bouts in fishing camps. She waits doggedly for him, continuing their work, resigned: "It doesn't surprise her that her marriage has failed to be an idyl of romance; she never expected it would be" (190). The marriage made by local "character" Uncle Stanley Gibb and "Miss Pauline" is a strained comedy. Sexton of the church and the town gravedigger, Uncle Stanley is a constant, garrulous witness to an empty house "where they live like strangers who happen to have rooms in the same hotel." He and his "Christian" wife share "a grim watchful armistice, likely to break out into hostilities any minute" (95).

The central tragedy of Jack Beechum's life, shown in greater detail in *The Memory of Old Jack,* is, likewise, his mismade marriage to Ruth Lightwood. Successful as a husband to his land and town, he fails in marriage. He and Ruth are very different people, and marriage proves "the great disaster of both their lives" (57). Strongly attracted to each other, they mislead themselves in illusion, and even their passions finally separate them, as "they lay beside each other in solitude, as rigid and open-eyed as effigies" (59). Turning to a secret love affair with the young widow Rose McGinnis, Jack is for a time delighted, but love needs the shape of a marriage to flourish in this world, and it is finally "as though he bore for these two women the two halves of an irreparably divided love. With Ruth, his work led to no good love. With Rose, his love led to no work" (134). Old Jack's unlucky story lingers in the mind of Port William, "troubling and consoling the night watches of lonely husbands and wives like a phrase from a forgotten song" (132).

The disrupted marriage begun by Virgil and Hannah Feltner has been quite the opposite, as clear-eyed and strong as there is in Berry's world, but it remains, ironically, for a bachelor, barber Jayber Crow, to give most eloquent voice to the husband's ideal state. He is a kind of "bachelor-witness" to the husband's life, monkishly eschewing marriage to celebrate it most intensely in higher form. In vowlike cadences, Jayber imagines it as

> a kind of last-ditch holy of holies: the possibility that two people might care for each other and know each other better than enemies, and better than strangers, and better than accidentally by happening to be alive at the same time in the same town; and that, with a man and a woman, this caring and knowing might be made by intention, and in the consciousness of all it is, and all it might be, and of all that threatens it. (97)

Crow's true name is Jonah, and he is a real survivor of his own trials, imagining in his cell above the shop, Port William as radiant, "a kind of Heavenly City, in which each house would be built in a marriage and around it, and all houses would be bound together in friendships, and friendliness would move and join among them like an open street" (97).

History, work, marriage—these are the major chords of Mat Feltner's life in his earthly city, and they resonate throughout *A Place on Earth*. Yet the husband's whole received and elected life is threatened by his son's death. Mat's chapter closes on the pain and darkness of its origin, and when he returns to earth, he faces and surpasses his crisis. The husband's task is to rediscover man's own mortal place, to yield his pride before a greater scheme, and to this end, he is instructed by his mystic and literal wives. Margaret speaks to him of endurance and acceptance of their mutual state: "From the day he was born I knew he would die.... I knew so well that he would die that when he did,

I was familiar with the pain. I'd had it in me all his life." She continues, confronting the central issue, "I don't believe that when his death is subtracted from his life, it leaves nothing." Mat is eased and renewed, her words fall "on his like light. . . . He feels rinsed and wrung, made fit" (451).

The husband is made ready for a second piece of advice, from his mystic wife, the soil. It is offered through Jack Beechum as he and Mat sit ritually through the night with Ernest Finley's body. Beechum once also lost his only son, and this is a tribal message from an ancient head to his successor, returning his attention to his roots in the earth.

> The old man spoke of names and landmarks of a time before Mat's birth, and Mat listened, his mind drawn back before its own beginning, held and quieted by the vision of another time, and by a sense of the quiet continuance of the land, the place, through all that has happened on it and to it—its troubling history of a little cherishing and much abuse. For as always it was finally the land that they spoke of, fascinated as they've been all their lives by what happened to it, their own ties to it, the wife of their race, more lovely and bountiful and kind than they usually have deserved, more severe and demanding than they have often been able to bear. (499)

Mat is comforted at the end, mainly by his rediscovery—for it is the husband's duty to find and find again such threads—of the process of renewal in nature. Mortal life is a compromise, and the splintered figures of man's stay, his atonement, are temporary. Sons, family, marriage, town—they will surely die (and Nathan will return from war, Hannah's child will be born, a new crop will be taken in, and younger husbands will step forward). Berry's husband to the world comes through with a richer, reconciling sense of the natural cycle. It is an old message, that "for every thing there is a season," as the preacher teaches in Ecclesiastes, and in Mat's daybook, he finally records coming to

> a vision of this land here underlying all the changes that come upon it and pass over it, all the lives that in their seasons rise on it to grow, bloom, make seed, die, and descend again into it. I've learned the pattern its lives make, from its men to its weeds, and I've grown ever less willing to set myself against any of them. (544)

"He must look to the woods," Wendell Berry advises us of the proper husband,[9] and Mat Feltner does just that. Walking his land, Mat is surprised to find a stand of trees overgrowing ruined tobacco plots, a healing of the "virgin wealth" his ancestors have spent. He sits down as leaves fall and night advances, and he "seems to come deeper into the presence of the place," to become "part of a design where death can give only to life." No longer struggling, he accepts his condition, "that the order he has made and kept in those clearings will be overthrown, and the effortless order of the wilderness will return to them" (550). He is eased. He is returning, once again, to his place on earth.

NOTES

1. *A Place on Earth* (New York: Harcourt Brace Jovanovich, 1967), 544. Subsequent page references are parenthetical in the text.

2. See especially the essays "A Secular Pilgrimage," "A Home to Dr. Williams," and "Discipline and Hope," in *A Continuous Harmony* (New York: Harcourt Brace Jovanovich, 1972).

3. "The Kentucky River: July, 1773," in *The Kentucky River: Two Poems* (Monterey, KY: Larkspur, 1975).

4. "Discipline and Hope," in *A Continuous Harmony*, 161. Subsequent page references are parenthetical in the text.

5. "Think Little," in *A Continuous Harmony*, 85.

6. *Nathan Coulter* (Boston: Houghton Mifflin, 1960), 18. Subsequent page references are parenthetical in the text.

7. *The Memory of Old Jack* (New York: Harcourt Brace Jovanovich, 1974), 215. Subsequent page references are parenthetical in the text.

8. The imagery is alive throughout Berry's work. The controlling metaphor of the essays in *The Hidden Wound*, for example, is the legacy of American racism as an ancient, unhealed injury.

9. "Discipline and Hope," in *A Continuous Harmony*, 161.

Annie Dillard
1945–

> That it's rough out there and chancy is no surprise. Every live thing is a survivor of a kind of extended emergency bivouac.
>
> —Annie Dillard, *Pilgrim at Tinker Creek*

One of Annie Dillard's distinguishing characteristics is that she is an intensely focused noticer of the natural world that surrounds her. She brings to her subject receptivity and readiness for wonder, as she explores nature's intricacies. Although her observations of the natural world are based on scientific understanding, her work is more notable for its metaphysical inquiries and conclusions.

Dillard was born on 30 April 1945, in Pittsburgh, Pennsylvania. Despite the vast urban environs of her childhood, she discovered pockets of nature hidden within the city, and a much wider presentation of the natural world on the shelves of libraries. "Everywhere, things snagged me," she relates in *An American Childhood.* She adds, "the visible world turned me curious to books; the books propelled me reeling back to the real world." Dillard received a B.A. (1967) and an M.A. (1968) from Hollins College, and she married Hollins English professor and writer Richard Dillard in 1965.

Dillard's education and observations came to fruition with the publication of her first book, *Pilgrim at Tinker Creek,* in 1974. The book received widespread critical acclaim, including the 1974 Pulitzer Prize for general nonfiction. *Pilgrim* has been compared with none less than Thoreau's *Walden,* and for good reasons. Both works are based on a period of solitary communion with nature, resulting in what Thoreau called "a meteorological journal of the mind." Dillard has described herself as "a poet, a walker with a background in theology and a penchant for quirky facts"; the same could be said of Thoreau.

Dillard's subsequent books, including *Tickets for a Prayer Wheel*

(1974) and *Holy the Firm* (1977), are also characterized by her intensity of seeing, her alertness to detail, and her sense of wonder. Yet despite her literary achievements, Dillard has been a source of frustration for environmentalists, because of her unwillingness to speak directly to ecological issues. "Here is our world as I see it," she seems to say. "Come along, or not. I shall be here in any matter."

"Pray Without Ceasing": Annie Dillard among the Nature Writers

James I. McClintock

"Sons and daughters of Thoreau abound in contemporary American writing," Edward Abbey writes in his introduction to *Abbey's Road* (1979), mentioning Edward Hoagland, Joseph Wood Krutch, Wendell Berry, John McPhee, Ann Zwinger, and Peter Matthiessen, as well as himself.[1] He reserves his highest praise for Annie Dillard, who "is the true heir of the Master." The others are Thoreauvian primarily in their identification with special locales—from Central Park in Hoagland's essay to Zwinger's Rockies. Abbey's one objection to Dillard's "otherwise strong, radiant book [*A Pilgrim at Tinker Creek*] is the constant name dropping. Always of one name"—God (*Abbey's Road*, xx). Abbey's assessment is astute, because it highlights the essential characteristics of Annie Dillard's nature writing: her writing about place, the language she uses to evoke her experiences, and her religious preoccupation and vocation. Abbey's assessment is also eccentric, because his objection to her religious preoccupation is directed at Dillard's most distinctive achievements in the nature essays of *A Pilgrim at Tinker Creek* (1974). The objection would apply also to *Holy the Firm* (1977) and *Teaching a Stone to Talk* (1982), the two other Dillard books that use nature as a touchstone for spiritual insight.[2]

Nature writing in America has always been religious or quasi-religious. All the important studies on the subgenre conclude that nature writing is "in the end concerned not only with fact but with fundamental spiritual and aesthetic truth."[3] That is true of essays by Thoreau, John Muir, John Burroughs, Aldo Leopold, Edwin Teale, and Joseph Wood Krutch, whose works represent more than a century of American nature writing. And Edward Abbey's work is infused with spiritual impulse, as he engages "Mystery."[4]

I suspect that Abbey's objection to Dillard's name-dropping is that her God is identifiably Judeo-Christian. That objection is understandable, because nature writers and, more broadly, conservationists, environmentalists, and students of American responses to nature have consistently held the Judeo-Christian tradition responsible for land abuse. In *A Sand County Almanac* (1949), for example, conservationist and nature essayist Aldo Leopold objected to an "Abrahamic concept of land" as commodity for technological man's use.[5] Historian Lynn White, Jr., concluded that the root of the postwar ecological crisis is a Judeo-Christian tradition that desacralizes nature and gives man "dominion" over it.[6] Rejecting Judeo-Christian anthrocentricity, writers have turned to spiritual alternatives. In "Lord Man: The Religion of Conservation," Steven Fox identifies many nature writers and conservationists who "embraced a variety of non-Christian religions."[7] Typical of many, poet and environmental activist Gary Snyder embraced Zen Buddhism and drew from Native American religious spiritual practices. Others, such as Joseph Wood Krutch, rejected Christian orthodoxy at first and a stoical humanism later, to embrace, finally, a pantheism that gave Krutch the profound sense that "we are all in this together," and that thus mirrored the thought of photographer Ansel Adams, who described his spiritual perspective simply as "a vast impersonal pantheism."[8] A pantheistic perspective fits well with the insights of modern ecological science, as is seen in Aldo Leopold's essay "Thinking Like a Mountain," an account of his "conversion" from an anthropocentric to a biocentric stance.[9]

Nature writing, then, has been broadly religious in the sense that Wendell Berry finds religion in the poetry he most highly values—poetry that has a "sense of the presence of mystery or divinity in the world" and "attitudes of wonder or awe or humility before the works of the creation." Such poets, like nature writers, go on what Berry calls "a secular pilgrimage," which "seeks the world of the creation, the created world in which the Creator, the formative and quickening spirit, is still immanent and at work."[10]

Theology has always attracted Annie Dillard. As an adolescent attending a Presbyterian summer vacation bible camp, she realized: "I had a head for religious ideas. They were the first ideas I ever encountered. They made other ideas seem mean."[11] *A Pilgrim at Tinker Creek, Holy the Firm,* and *Teaching a Stone to Talk* are saturated with religious thought, longing, and experience. Dillard is after the "pearl of great price," religious vision, which will reconcile the self—which is pulled between faith and doubt—with a nature that is often cruel and ugly, and with a God who seems as irrational as loving. She is within meditative traditions and records repeated mystical experiences. She prepares for mystical reconciliation by performing rituals that mingle conventions for encountering nature that are found in nature writing with Judeo-Christian traditions and rituals. She is an offbeat Christian who walks in nature and reads science as part of her preparation for vision.

In *Holy the Firm*, Dillard finds unsatisfactory the "accessible and universal view," mentioned by Wendell Berry and "held by (Meister) Eckhart and by many peoples in various forms, ... that the world is immanation, that God is in the thing, and eternally present here if nowhere else" (73–74). That view is "scarcely different from pantheism," she writes, because from that perspective, "Christ is redundant and all things are one" (73–74). This statement sets Dillard apart from the other nature writers; her perspective is Christian.

Dillard's books are dotted with biblical allusions, and she unself-consciously uses the word *Christ*. During the central mystical moment in *Holy the Firm,* to cite the most extended example, she is walking home from a country store with communion wine for her church when suddenly she is filled with light, "everything in the world is translucent," the bay below is "transfigured," "everything is whole, and a parcel of everything else," and she sees that "Christ is being baptized" by John (68–71). Christ "lifts from the water. Water beads on his shoulders. I see the water in balls as heavy as planets, a billion beads of water as weighty as worlds, and he lifts them up on his back as he rises" (70). Dillard writes throughout *Holy the Firm* in imagery that evokes the opening of the book, when she wakens, looks across Puget Sound, and greets the morning:

> I wake in a god. ... Someone is kissing me—already. ... I open my eyes. The god lifts from the water. His head fills the bay. He is Puget Sound, the Pacific; his breast rises from pastures; his fingers are firs; islands slide wet down his shoulders. Islands slip blue from his shoulders and glide over the water, the empty, lighted water like a stage. (4)

In fact the entire structure of *Holy the Firm* is Christian. Dillard equates the three days the book spans to Creation, the Fall, and Redemption; and the subjective framework is "the tripartite pattern of faith, doubt, and faith renewed."[12] Robert Dunn notes that the book's three chapters parallel the "three stages of the mystic way—illumination, purgation, and union."[13]

Pilgrim at Tinker Creek also opens with Dillard awakening to a world seen through Christian experience, even if her doubt is constant. She is, after all, an anchorite and a pilgrim, awakened in the morning to the possibility of mystery by her cat, which has left her "body covered with paw prints in blood: I looked as though I'd been painted with roses" (1-3). This imagery is profoundly linked to the Judeo-Christian tradition through the Passover, on the one hand, and through Christ's redemptive blood and the rose symbolizing Mary, on the other. The central mystical experience in *Pilgrim at Tinker Creek,* the vision of "the tree with the lights in it," is a revelation of "Christ's incarnation," which Dillard accepts despite liberal theological objections to a belief that Christ's incarnation took place at a particular time and a particular place.[14] She affirms "the scandal of particularity," because "I never saw a tree that was no tree in particular" (81); the tree with lights on it is, after all, a particular backyard cedar.

Though Annie Dillard sees from the standpoint of Christian orthodoxy, she is still heterodox and unconventional. Critic Margaret Reimer has shown that though "Dillard stands in the orthodox Christian tradition" in her views of evil, for example, "her conclusions (or the lack of them) are far from the traditional Christian answers."[15] Dillard has always been uncomfortable within orthodoxy, even though, paradoxically, she is also uncomfortable outside a Christian perspective. From childhood on, she was neither quite inside nor completely outside conventional religious experience. When she went off to a Presbyterian summer bible camp where she learned she had a "head for religious ideas" and got "miles of Bible by memory," she was aware that her parents would have objected to the evangelical intensity of "the faith-filled theology ... only half a step out of a tent" (132-33). As an adolescent, she was already absorbed in the theological question that is at the center of both *Pilgrim at Tinker Creek* and *Holy the Firm*—"If the all-powerful Creator directs the world, then why all this suffering?" She had written a paper about the Book of Job, but she had also quit the Presbyterian church. Her off-tempo relation to Christianity is caught in the moment when she meets with her family's minister to tell him of her decision to quit but, at the same time, accepts from him books by C. S. Lewis, including *The Problem of Pain* (*AC,* 227–28). More than two decades later, she still has not found her institutional place, although she attends church. In *Teaching a Stone to Talk,* she notes that she has "overcome a fiercely anti-Catholic upbringing in order to attend Mass"; but she does so "simply and solely to escape Protestant guitars" and likens her attendance to having "run away from home and joined the circus as a dancing bear" (18–19).

At times, Dillard strains to remain Christian. For instance, she rejects pantheistic immanence—that "God is in the thing," referred to above—but cannot quite accept the conventional Christian view that "emanating from God, and linked to him by Christ, the work is infinitely other than God" (*HF,* 73). While the concept of emanance permits a representation of Christ that allows for the salvation of "the souls of men," it leaves the rest of nature "irrelevant and nonparticipant," unreal to "time," "unknowable, an illusory, absurd, accidental, and overelaborate state"—fallen, in a word (*HF,* 74). Unwilling to accept a view that denies a sacralized, familiar natural world, Dillard entertains a view from "esoteric Christianity" that there is a substance called "Holy the Firm" that is "in touch" with both the lowest of material reality—the "salts and earths"—and the absolute (*HF,* 72). The absolute and the most ordinary aspects of nature are connected: "Matter and spirit are of a piece but distinguishable; God has a stake guaranteed in all the world" (*HF,* 75). Characteristically, affirmations are undercut, this time with the anticlimactic aside that "these are only ideas" (*HF,* 75). For Dillard, however, there is no such thing as "only" ideas. She proves herself outside orthodoxy and beyond conventional Christianity, without abandoning Christian preoccupations, beliefs, and longings.

As Reimer has shown, Dillard's "theology is always dialectical" and contains "both the conventional language of religious mysticism as well as more macabre elements of religious experience" (187). The dialectical tension is between "the material and the spiritual, the natural and the transcendent . . . the beauty and the horror within the natural world" (182). I agree with Reimer's assessment that "the power of Dillard's vision arises from her strength to maintain the contradictions within a single vision" (189). Dillard's vision is contradictory at its most extreme, and dialectical in its most powerful insights. The kinds of ritual she creates and writes about explain in large measure how she balances these unresolved contradictions within a single, unified vision. Her rituals are familiar to both religious practitioners and nature observers.

Students of myth and ritual know that worldviews, or myths, contain contradictions and unresolved mysteries that adherents live with despite doubt, and that ritual is a way both of moving toward deeper understanding and of affirming belief publicly—a way of acting, without complete knowledge. Annie Dillard seeks a vision that is the "pearl of great price," which "may be found" but "may not be sought" (*PTC,* 342), so the question becomes "how then is she to act? How is the search to be conducted?"[16] Annie Dillard's ritual acts allow her to affirm life and God without a theological resolution of fundamental religious questions. Through these rituals, she strives for—and experiences—reconciliation between herself, a sometimes horrible—as well as beautiful—nature, and a mysterious God who, at times, seems as maniacal as loving. Fittingly, the rituals are a blend of Judeo-Christian rites in nature; they are the rituals of stalking, seeing, and dancing.

Walking, as more than exercise, has a long tradition in literature, from Plato's walks when he formulated his dialogues, to Saint Augustine's walk on the seashore, to the walks of seventeenth-century Christian literary walkers: "The walk is an occasion and setting for revelation, for a sudden increase in their awareness of the indwelling of God in the world."[17] Walkers are pilgrims seeking visions. As Thoreau comments in "Walking," those few who understand "the art of Walking," who "have a genius for *sauntering,*" are linked with medieval pilgrims about whom children exclaimed, "'There goes a-*Sainte-Terre,*' a Saunterer, a Holy-Lander."[18] Those who walk in Thoreau's way

> saunter toward the Holy Land, till one day the sun shall shine more brightly than ever he has done, shall perchance shine into our minds and hearts, and light up our whole lives with a great awakening light, as warm and serene and golden as on a bankside in autumn. ("Walking," 136)

Thoreau's imagery of light is echoed in Dillard's mystical moments, as it is in all the mystical tradition, including the Christian. John Elder writes that for inveterate walker William Wordsworth, the "Pilgrim" of "The Prelude," "walking is a process of reconciliation: it provides the dynamic unity of his life" and art.[19] "The Prelude," for example, is a work organized in part by walking. Elder, in ways applicable to Dillard's essays, writes about walking in the works of others, such as contemporary poet A. R. Ammons. That is particularly true if we remember that Dillard's vision is dialectical. Writing about Ammons, Elder might as well be writing about Dillard: "There is no absolute unity available for existence in a physical, and thus temporal, work. Rather, going from one foot to the other, human life takes its passage through a universe of particulars"; and the major response to the relations between nature, human imagination, and spirit is "one of ambivalence: right foot, left foot" (99, 100).

In the chapter "Stalking" in *Pilgrim at Tinker Creek,* Dillard tells us she learned to stalk fish and muskrats, who "by their very mystery and hiddenness crystallize the quality of my summer life at the creek" (188). Learning to stalk muskrats took "several years," until one evening, when she had "lost" herself, "lost the creek, the day, lost everything but (the creek's) amber depth," a young muskrat "appeared on top of the water, floating on its back" (*PTC,* 194). She was ecstatic. The excitement and wonder of sighting an "ordinary" muskrat through her ritual stalking is described with the language of revelation. She records her joy and surprise "at having the light come on so suddenly, and at having my consciousness returned to me all at once and bearing (a) . . . muskrat" (194). Fearing that the encounter was a once in a lifetime experience, she stalks muskrats day and night; and at the point she sees another, she reports, with the Thoreauvian extravagance that Edward Abbey so admired, "My life changed" (195). What Dillard calls stalking is, obviously, closer to meditating. The *"via negative,"* she says, is a form of stalking "as fruitful as actual pursuit" (187). She waits "emptied," like "Newton under the apple tree, Buddha under the bo" (187). Dillard reminds us that Ezekiel "excoriates" false prophets who will not go up into the gaps, and she exhorts us to "stalk the gaps," which are the cliffs on the rock where you "cower to see the back parts of God" (276). Such stalking will reveal "more than a maple," she writes; it will reveal "a universe" (276).

Dillard's walks around Tinker Creek in Virginia and her stalking of the muskrat reveal not merely the habits of the secretive animal, for the mystery and hiddenness she often attributes to muskrats are those she most often attributes to God. Moreover, her personal ritual of stalking is ultimately described in Christian terms: on the night her life changed as a result of seeing the muskrat, she summarizes the nature of the stalking ritual as "Knock; seek; ask," obviously a variant of the biblical "Ask, and it shall be given to you; seek, and ye shall find; knock, and it shall be opened to you. For everyone that asketh receiveth; and he that seeketh findeth; and to him that knocketh it shall be opened" (195). In Dillard's work the nature-writing conventions of encountering nature directly and immediately through such ordinary activities as walking while one is open to aesthetic and spiritual experience mingle and meld with Christian ritual, tradition, and experience. Ordinary experince fuses with the millennial, the temporal with the transcendent.

In *Holy the Firm* and *Teaching a Stone to Talk,* Dillard is more conventional in her use of the walking ritual than in *Pilgrim at Tinker's Creek,* but she makes the same points. In *Holy the Firm,* she is deeply troubled by the terrible suffering of Julie Norwich, who is in the hospital, her face burned in a plane accident. Worrying about and questioning the Christian response to the sufferings of the innocent, Dillard, near despair, asks, "Do we really need more victims to remind us that we're all victims?" and she reminds herself that we are "sojourners in a land we did not make, a land with no meaning of itself and no meaning we can make for it alone" (*HF,* 62, 63). In this state of mind, she feels unworthy to buy the communion wine she had volunteered to get, but she goes anyway. She walks home, "and I'm on the road again walking, my right hand forgetting my left. I'm out on the road again walking, my right hand forgetting my left. I'm out on the road again walking, and toting a backload of God" (66). As she starts up a hill, the landscape starts "to utter its infinite particulars," and she lists particular features of the landscape about her—"blackberry brambles, white snowberries, red rose hips, gaunt and clattering broom" (67). Soon, the particulars are alive: "mountains are raw nerves;... the trees, the grass... are living petals of mind." Finally,

> walking faster and faster, weightless, I feel the wine. It sheds light in slats through my rib cage, and fills the buttressed vaults of my ribs with light pooled and buoyant. I am moth; I am light. I am prayer and I can hardly see. (68).

At that moment, she experiences the vision that is central to the book; she beholds Christ being baptized.

The essays in *Teaching a Stone to Talk* often expand the notion of ordinary walking to larger journeys and expeditions. In "Sojourner" she notes that the title word appears frequently in the Old Testament and "invokes a nomadic people's sense of vagrancy, a praying people's knowledge of estrangement, a thinking people's intuition of sharp loss" (*TST,* 150). Thus, she alternates, in this essay and in her writing in general, between "thinking of the planet as time" and "as a hard land of exile in which we are all sojourners" (150). A number of the essays in *Teaching a Stone to Talk* explore the dialectic between being at home and being estranged, as she moves her setting from Tinker Creek to places as remote as the Napo River in the Ecuadorian jungle and the North Pole. In "An Expedition to the Pole," she combines personal experience, history, and fantasy. The personal experience of visiting the Arctic and viewing the Arctic Sea fuses with the history of various Polar expeditions that entailed enormous suffering for ill-equipped explorers. She fantasizes that she has "quit my ship and set out on foot over the polar ice," and that she has traveled across an ice floe, where she encounters both historical personages and members of the congregation of the Catholic church she has been attending. They are all together on a spiritual quest. Her attendance at Catholic services is part of her search for "the Pole of Relative inaccessibility," or "The Absolute" (19). She asks, "How often have I mounted this same expedition, has my absurd barque set out half-caulked for the Pole?" And she quotes Pope Gregory in seeking to define her aim: "'To attain to somewhat of the unencompassed light, by stealth'" (44). Although Dillard emphasizes alienated experience because she is poorly and absurdly equipped for the spiritual expedition to the Pole, she ends the essay with a fantasy in which she is on the floor with the church members, "banging on a tambourine" and singing loudly. "How can any of us tone it down?" she asks, "for we are nearing the Pole" (52). Dillard actually seeks and creates the conditions for ecstatic, mystical experience; doubt and hope are held in balance within the imaginative framework of sojourning, of exploring on foot.

Annie Dillard walks and stalks so that she can "see" in more than one sense. To see truly, she must prepare herself ritualistically, must become both innocent and informed. Dillard defines innocence as "the spirit's unselfconscious state at any moment of pure devotion to any object. It is at once receptiveness and total concentration" (*PTC*, 83). Innocence is a state she values as highly as did the romantics and Christians before her. That may be why she often identifies the stalking and seeing rituals with childhood and childhood games. "Only children keep their eyes open," she writes (*PTC,* 104). She describes nature as "like one of those line drawings of a tree that are puzzles for children: Can you find hidden in the leaves a duck, a house, a boy, a bucket, a zebra, and a boot?" (*PTC,* 18). The universe is a merry-go-round, and the cost is a child's rubber duck (*PTC,* 23, 45). Dillard evokes her childhood, as well as others', and always in the service of seeing, in all senses of the word, the microcosm of Tinker Creek: "If I seek the senses and skill of children ... I do so only, solely, and entirely that I might look well into the creek" (*PTC,* 104). A major motif of the book is, as I have already noted, the hiddenness of God as well as of nature. Childhood games are played to coax the Creator from hiding. She alludes to John Knoepfle's poem in which "'christ is red rover ... and the children are calling / come over come over'" (*PTC,* 209). Longing for God, she compares the banging of her will against rock with a child beating on a door and calling: "Come on out! ... I know you're there" (209).

That she seeks to see by entering "the spirit's unselfconscious state" through "pure devotion to any object" is one of many obvious signs that Annie Dillard is intensely aware of her absorption in meditative traditions. Her efforts to see are rewarded in the numerous mystical moments recorded in *Pilgrim at Tinker Creek, Holy the Firm,* and *Teaching a Stone to Talk.* One summer evening, when she is practicing being "an unscrupulous observer" of shiners feeding in Tinker Creek,

> something broke and something opened. I filled up like a new wineskin. I breathed an air like light; I saw a light like water. I was the lip of a fountain the creek filled forever; I was ether, the leaf in the zephyr; I was flesh-flake, feather, bone. (*PTC,* 34–35).

Because of such moments, critics have rightly seen in Dillard's mystical experiences parallels with Ralph Waldo Emerson's experiences and views recorded in "Nature." Dillard's observation that "there is [a] kind of seeing that involves a letting go [and] when I see this way I sway transfixed and emptied" is justifiably compared with Emerson's famous statement that "I became a transparent eyeball; I am nothing; I see all; the currents of the Universal Being circulate through me."[20]

Dillard also is Emersonian in preparing herself for vision by exercising her "Understanding," by disciplining her nature experiences wth scientific information and ideas. While she rightly states, "I am no scientist" (*PTC,* 12), her essays are packed with allusions to scientific reading of all sorts. Not surprisingly, those allusions fall into two, dichotomous categories. One evokes a nature that is deterministic—the insect world, in which a giant water bug sucks out the innards of a frog, "a monstrous and horrifying thing" that leaves her deeply shaken (*PTC,* 6). In a more lighthearted moment, she makes the same point in a chapter about nature's horrors: "Fish gotta swim and bird [sic] gotta fly; insects, it seems, gotta do one horrible thing after another" (64). The other category of scientific allusions focuses on the indeterminant nature described by twentieth-century physics. In the chapter "Stalking," Dillard has a two-page commentary on Werner Heisenberg's "Principle of Indeterminancy," and she quotes, in addition, physicists Sir Arthur Eddington and Sir James Jeans, whose views, she notes gleefully, mean that "some physicists now are a bunch of wild-eyed, raving mystics" (*PTC,* 206–8). To illustrate, she quotes Eddington's statement that the Principle of Indeterminancy "'leaves us with no clear distinction between the Natural and the Supernatural'" (207–8).

Critics, especially Gary McIlroy and Margaret Reimer, have done very well in pointing out Dillard's response to science, especially to the Principle of Indeterminancy.[21] Her ritual preparation for seeing, however, has depended on a broader range of science and science-related reading than has been discussed. Ofter her reading is specific to phenomena she observes. When she stalks the muskrat, she refers to biologist and expert on muskrats Paul Errington (197). She refers to biologist and science historian Howard Ensign Evans on dragonflies (67), limnologist Robert E. Coker on plankton movement, Rutherford Platt on trees (noting that his *The Great American Forest* is "one of the most interesting books ever written" [97]), and so on. In Emersonian fashion, she "disciplines" her "Understanding" in preparing for visions.

Two writers important in disciplining Dillard's understanding are Frenchman Henri Fabre and American Edwin Way Teale, sources for a number of her comments on the horrors of the insect world. In her chapter "The Fixed" in *Pilgrim at Tinker Creek,* which contains some of her most pessimistic and horrific conclusions, Dillard refers frequently to turn-of-the-century Fabre. She notes that "even a hardened entomologist like F. Henri Fabre confessed to being startled witless every time" a praying mantis strikes its prey, and she quotes a long passage of his describing the macabre mantis mating ritual during which the female gnaws on her "swain" until there is just that "masculine stump" going "on with the business" (*PTC,* 56, 59).[22] Edwin Way Teale is the most frequently cited writer in the other dark chapter in *Pilgrim at Tinker Creek,* "Fecundity," in which nature seems primarily a matter of eating, breeding, and dying: the "universe that suckled us is a monster that does not care if we live or die.... It is fixed and blind, programmed to kill" (180). She illustrates this grim, amoral natural world with examples drawn from Teale's *The Strange Lives of Familiar Insects,* which is, she exclaims, a "book I couldn't live without" (171).

Although Dillard draws on Fabre's and Teale's writings to underscore a deterministic, amoral natural world that may be "the brainchild of a deranged manic-depressive with unlimited capital," these writers achieved their fame as popularizers of science by maintaining optimistic spiritual outlooks (*PTC,* 67). Fabre never accepted Darwinian evolutionary theory and remained a devout Roman Catholic. A humble French provincial who was not accepted by the academy until very late in life, Fabre was less the laboratory scientist in a white lab coat than a living example of the persona familiar to the nature essay in general and to Annie Dillard's essay in particular—the amateur who is faithful to his local environment and who experiences awe and wonder in nature's small moments. Edwin Way Teale, who admired Fabre and introduced the English translation of his collected essays, is also optimistic, despite his chronicles of the violent and grotesque insect world.[23] In *Speaking for Nature,* Paul Brooks describes Teale as one of the finest "literary naturalists" since Thoreau. He adds that Teale, with others, has "opened the eyes of millions of readers... to a widespread feeling of kinship with the other forms of life with which we share the earth" (xiv). From his *Strange Lives of Familiar Insects* to his widely read and well-regarded books on the American seasons, Teale's works reflect the affirmation and joy characteristic of American nature writers and of one side of Annie Dillard's dialectical view. The two men offered her more than scientific information.

In her intellectual preparations for reaching a state of innocence followed by mystical insight, Dillard's reading is often as much the focus of her attention as the natural object itself. Some of the more important categories in her diverse reading are theology and other religious matters (Martin Buber, Thomas Merton, Julia[n] of Norwich, and the Koran); art (DaVinci, Van Gogh, Breughel, and El Greco); adventure (Lewis and Clark, Heyerdahl, and the Franklin Polar expedition); and literature (Thoreau, Coleridge, Blake, Goethe, and Eliot). A full accounting of the interplay between her reading and her responses to nature is impractical here, but one intellectual source is crucial—the philosopher Heraclitus.

Dillard associates Heraclitus with views close to those of quantum physics, that "nature is wont to hide herself" (205). Moreover, his perspective is akin to her dialectical vision. Dillard opens *Pilgrim at Tinker Creek* with the following epigraph from Heraclitus:

It ever was, and is, and shall be,
ever-living Fire, in measures being
kindled and in measures going out.

Heraclitus was the philosopher of opposites. But they are opposites that have underlying connections; for instance, good and evil define one another. The same is true for all natural events. Though they are described and seen in terms of opposites, there is an underlying interrelatedness, a hidden connection, of which fire is the physical embodiment.[24] The epigraph in *Pilgrim at Tinker Creek* and the pervasive fire imagery in *Holy the Firm* are signs that the intellectual aspects of Dillard's meditations prepare her for a sense of wonder no less than horrific vision. As a result, she sees not only the dead frog but also the "tree with the lights in it ... transfigured, each cell buzzing with flame" (*PTC,* 35). It is a vision that, as Heraclitus would have predicted, comes and goes. It is a vision she lives for. In that moment, her spirit's aspirations and her own reality are confirmed. In *Holy the Firm,* the fire is the fire that attracts the moth to destruction and the fire that disfigures Julie; but it is also the light that comes into Dillard's spirit and onto her face, as onto the face of every artist, which, "like a seraph's" face, lights "the kingdom of God for the people to see" (77). Heraclitus's imagery of forever waxing and waning fire is the perfect metaphor for her thematic dualities for good and evil, beautiful and grotesque, and repulsive and awesome—all of which coexist in God's nature.

At times, however, Dillard's ritual stalking and ritual preparations to see are undercut by nature's grotesquery. Reading, in particular, is not a suffucient stay against confusion. Dillard discovers that she can get lost in the "labrinthine tracks of the mind," when she most needs to live in the senses: "So long as I stay in my thoughts ... my foot slides" and "I fall" (*PTC,* 88). In a passage from *Teaching a Stone to Talk* that echoes the creekside frog episode in *Pilgrim at Tinker Creek,* Dillard encounters a resting Guernsey cow during one of her rambles but is shocked to find that it is dead. The cow's insides are gone, "her udder and belly ... open and empty." Horrified, Dillard sees that the cow's legs had broken when a limestone sinkhole had suddenly opened under her weight (*TST,* 168). Dillard is shocked and disoriented, fearing that the ground will open beneath her and she will fall unchecked (168). The alternative to falling, to terror, and to doubt, she writes, is to dance (*PTC,* 88). Ritual dance, real or imagined, allows Dillard to quiet morbid intellectualizing, to enter into direct contact with the natural world, and to praise despite the threat of meaninglessness. Dance is her least-mentioned ritual, but it is crucial for keeping her spiritual balance.

In *Pilgrim at Tinker Creek,* falling and dancing imagery combine as Dillard seeks signs of hope but fears that the monstrous may prevail. Near the book's end, she describes herself as a "sojourner seeking signs" and remembers that Isak Dinesen, brokenhearted, had stepped into the Kenyan morning seeking a sign and had witnessed a rooster tear from its root a chameleon's tongue—the unwelcome sign again of pervasive cruelty. But Dillard's thoughts and feelings about that shocking moment in Dinesen's experience and about cruelty in nature are altered; she is once again "transfigured" as a maple key twirls down toward her on the wind. She becomes aware of that other "wind of the spirit" and thinks, "If I am a maple leaf falling, at least I can twirl" (275–76). Similarly, in concluding "Sojourner" in *Teaching a Stone to Talk,* Dillard turns from "thoughts of despair" about purposelessness sensed everywhere, to thoughts about beauty, and she invites us, "with as much spirit as we can muster, [to] go out with a buck and wing" (152). That said, she envisions nature joining in:

> The consort of musicians strikes up, and we in the chorus stir and move and start twirling our hats. A Mangrove island turns drift to dance... rocking over the salt sea at random, rocking day and night and round the sun, rocking round the sun and out toward east of Hercules. (152)

Moreover, as she has with stalking and seeing, Dillard locates the dance ritual in the tradition of Judeo-Christian ritual and mysticism. She recalls that King David "leaped and danced naked before the dark of the Lord in a barren desert," a model for herself in the face of spiritual emptiness, and a reminder to us, she says, to "make connections; let rip; and dance whenever you can" (97–98). She is dancing significantly, as *A Pilgrim at Tinker Creek* ends: "I go my way and my left foot says 'Glory,' and my right foot says 'Amen' in and out of Shadow Creek, upstream and down, exultant, in a daze, dancing to the two silver trumpets of praise" (279).

There are powerful moments in *Pilgrim at Tinker Creek* when Annie Dillard, performing her stalking, seeing, and dancing rituals, has the sudden insight that not only is she stalking but she is being stalked, not only is she seeing but she is being seen, and not only is she dancing but music is being played for her. On the dark side, God is a stalker-hunter, a destroyer, the ultimate "'archer in cover,'" whose arrows bring fear and mortality (91). Being seen, though, is joyful. In the central mystical moment of the book, when Dillard is taken unaware by the "tree with the lights in it," she exclaims that "it was less like seeing than like being for the first time seen, knocked breathless by a powerful glance" (35). We do not need to be told who has seen her. And the agent of the dance is more than nature. When Dillard imagines herself spinning through the universe to stop her "sweeping fall," she notices that "Someone" pipes as "we are dancing a tarantella until the sweat pours" (23). Having divined that she is stalked, seen as well as seeing, a dancer to "Someone's" tune, Dillard concludes that she "cannot ask for more than to be so wholly acted upon," even if by a plague of locusts, because she would willingly pay the price in discomfort to be "rapt and enwrapped" in the "real world" (*PTC,* 226). Her imagery of being acted on climaxes as the book ends in passages about ritual sacrifice that, again, combine her personal vision with the Judeo-Christian.

As Dillard debates whether corruption and beauty are equal in creation and concludes that corruption is not "beauty's very heart," she describes herself as "a sacrifice bound with cords to the horns of the world's rock altar." There she takes a deep breath and opens her eyes, seeing "worms in the horn of the altar," as "a sense of the real exults me; the cords loose; I walk on my way" (*PTC,* 248). In this mystical moment, she finds freedom in accepting the fallen world. But she needs to go beyond supplication and acceptance to praise, from "please" to "thank you." And she does. The last two chapters of *Pilgrim at Tinker Creek* focus on sacrificial rituals that Dillard must know are in the Judeo-Christian tradition, rituals of purification and thanksgiving rather than merely propitiating the gods. She concentrates on an ancient Israelite ritual, "the wave breast of thanksgiving," which is—significantly, considering Dillard's joy in being seen—"a catching [of] God's eye" (266). The priest dresses in clean linen, comes to the altar, and is given a consecrated breastbone of a ritually slain ram, which he waves as an offering to the Lord. Dillard knows this ritual of thanksgiving works, and she ends her discussion with a phrase from Catholic liturgy: "Thanks be to God" (266). She then calls on a second part of the ritual to acknowledge her ongoing problem with the cruel, horrible, and monstrous in nature. After the priest waves the breastbone, he "heaves" the ram's shoulder bone. Dillard interprets this to mean that after catching

God's eye, one can "speak up for the creation," can protest cruelty and waste in the natural world. "Could I heave a little shred of frog shoulder at the Lord?" she muses, remembering the frog's death she had witnessed at creekside. She finally understands, though, that both the "wave" to capture God's glance and the "heave" to lodge protest are necessary for a unified ritual; "both meant a wide-eyed and keen-eyed thanks," and neither was whole without the other (279).

As one stalking and being stalked, seeing and being seen, dancing to someone's tune, performing rituals of sacrifice, and serving as victim of the sacrifice, Dillard places herself in a mystical relationship with both a nature and a God who are at once both concealed and revealed. She concludes with prayers of affirmation, no matter how bleak the moment's reality is. In *Teaching a Stone to Talk*, she notes that "we as a people have moved from pantheism to pan-atheism" (69). We have desacralized nature, she says, and God no longer speaks from the whirlwind. Until "God changes his mind, until the pagan gods slip back to their hilltop groves, or until we can teach a stone to talk, all we can do with the whole inhuman array is watch it" (72). Dillard observes: "we are here to witness.... The silence is all there is." Nevertheless, she concludes with an exhortation to prayer: "you take a step in the right direction to pray to this silence.... Pray without ceasing" (72, 76).

The remarkable conclusion to *Holy the Firm* is, in effect, an extended, unceasing prayer that reveals the God beyond nature who is linked with it by "holy the firm." Here the book's major thematic and artistic elements coalesce. The images of the burning moth, Julie the burned child, and Annie Dillard herself are intertwined with Puget Sound's "islands on fire" and seraphs' and the artists' faces that "can sing only the first 'Holy' before the intensity of their love ignites them again and dissolves them again, perpetually, into flames" (*HF*, 45; Scheick, 58). Dillard is the artist-nun, aflame with holiness. Dillard's book-prayer is "lighting the kingdom of god for the people to see" (*HF*, 77).

If Dillard's Christian desire to light the kingdom of God marks her apart from other nature writers, it is only a matter of degree. All non-Christian writers I have mentioned, and many more, are fascinated with the relationships between nature, human consciousness, and "mystery." Troubled by the combined intellectual and spiritual consciousness of Newtonian and Cartesian thought, which separated spirit and matter and placed nature beneath humans in importance, by nonteleological, Darwinian natural selection, and by technological assaults on the natural environment, twentieth-century nature writers have explored alternative views. Committed to science as guide, writers as diverse as Joseph Wood Krutch, Edward Abbey, Loren Eiseley, Peter Matthiesen, Barry Lopez, and Ann Zinger have nevertheless kept as their first loyalty and touchstone direct, experiential encounters with nature. All report aesthetic and spiritual rewards for doing so. In Dillard's essays, the same persona speaks to us as from the works of other nature writers—the solitary figure in nature, moved to philosophical speculation and, finally, to awe and wonder, to self-forgetting, and to an affirmation of realities that resist modern and contemporary threats of hopelessness and despair.

Despite such affinities, however, Annie Dillard has a special voice that speaks of balancing the tension between fear and hope, between horror and celebration, through rituals of stalking, seeing, and dancing. In such rituals, she has awakened not only to mystery in nature but to mystery beyond nature. Her Christian obsessions and ritual practice culminate in prayer without cessation. "[I] resound," she writes, "like a beaten bell" (*PTC*, 13).

NOTES

1. *Abbey's Road* (New York: E. P. Dutton, 1979), xx.

2. *Pilgrim at Tinker Creek* (New York: Bantam Books, 1974); *Holy the Firm* (New York: Bantam Books, 1977); *Teaching A Stone to Talk: Expeditions and Encounters* (Cambridge: Harper and Row, 1982). Cited in text as *PTC, HF, TST.* Born in 1945, Dillard received B.A. and M.A. degrees from Hollins College, was a contributing editor to *Harper's Magazine*, and has taught creative writing at various colleges and universities. She received a Pulitzer Prize in 1974 for *Pilgrim at Tinker Creek*. Her books on other subjects are *Tickets for a Prayer Wheel* (Columbia: University of Missouri Press, 1974); *Living By Fiction* (New York: Harper and Row, 1982); *Encounters with Chinese Writers* (Middletown: Wesleyan University Press, 1984); *An American Childhood* (New York: Harper and Row, 1987); and *The Writing Life* (New York: Harper and Row, 1989).

3. John Hildebidle, *Thoreau: A Naturalist's Liberty* (Cambridge: Harvard University Press, 1983), 61.

4. Abbey's most direct comment on "Mystery . . . with an emphatically capital M" occurs in "The Ancient Dust," in *Beyond the Wall: Essays from the Outside* (New York: Holt, Reinhart and Winston, 1984), 154.

5. *A Sand County Almanac with Essays on Conservation from Round River* (New York: Ballantine Books, 1968), xviii.

6. "The Historical Roots of Our Ecological Crisis," *Science* 155 (1967): 1203–7.

7. *John Muir and His Legacy: The American Conservation Movement* (Boston: Little, Brown and Company, 1981), 363.

8. "April, The Day of the Peepers," in *The Twelve Seasons: A Perpetual Calendar for the Country* (New York: William Sloane, 1948), 13; Adams quoted in Fox, *John Muir and His Legacy*, 363.

9. *Sand County Almanac*, 137–41.

10. "A Secular Pilgrimage," in *A Continuous Harmony: Essays Cultural and Agricultural* (New York: Harcourt Brace Jovanovich, 1972), 66.

11. Dillard, An American Childhood (New York: Harper and Row, 1987), 133. Cited in text as *AC*.

12. Nancy Lucas, "Annie Dillard," in *Dictionary of Literary Biography Yearbook: 1980* (Detroit: Gale Research Co., 1981), 187; quoted in William J. Scheick, "Annie Dillard, Narrative Fringe," in *Contemporary American Women Writers: Narrative Strategies*, ed. Catherine Rainwater and William J. Scheick (University of Kentucky Press, 1985), 63: Scheick, "Annie Dillard," 61.

13. "The Artist as Nun: Theme, Tone and Vision in the Writings of Annie Dillard," *Studia Mystica* 1, no. 4 (1978): 18.

14. *PTC*, 81; see comments on the scandal of the particular in *HF*, 55–56.

15. "The Dialectical Vision of Annie Dillard's *Pilgrim at Tinker Creek*," *Critique* 24 (1983): 189.

16. David L. Lavery, "Noticer: The Visionary Art of Annie Dillard," *Massachusetts Review* 21 (1980): 257.

17. Tom Wolf, "A New Walk Is a New Walk," *Walking Magazine*, Autumn, 1986, 64–65.

18. Henry David Thoreau, *The Natural History Essays* (Salt Lake City: Perrigrine Smith, 1980), 93.

19. *Imagining the Earth: Poetry and the Vision of Nature* (Urbana: University of Illinois Press, 1985), 97.

20. Emerson (1836) quoted in John Conron, ed., *The American Landscape* (New York: Oxford University Press, 1974), 581.

21. See Gary McIlroy, "*Pilgrim at Tinker Creek* and the Burden of Science," *American Literature* 59, no. 1 (1987): 71–84; and Reimer, "The Dialectical Vision."

22. Dillard is unaware that this popular conception is no longer a scientific view.

23. Alexander Teixeira de Mattos, trans., *The Insect World of J. Henri Fabre* (New York: Dodd, Mead and Co., 1961).

24. "Heraclitus," *The New Encyclopaedia Britannica*, 15th ed. vol. 5 (Chicago: Encyclopaedia Britannica, 1988), 860.

Pilgrim at Tinker Creek and the Social Legacy of *Walden*

Gary McIlroy

Readers of Annie Dillard's *Pilgrim at Tinker Creek,* a work in many ways reminiscent of *Walden,* are usually disappointed by its virtual neglect of society. It is accomplished, says Hayden Carruth, "with little reference to life on this planet at this moment, its hazards and misdirections, and to this extent it is a dangerous book, literally a subversive book, in spite of its directions."[1] This must be put into perspective. *Pilgrim at Tinker Creek* has been well received. It won the Pulitzer Prize for general nonfiction and has been anthologized in over thirty prose collections. Many critics, nevertheless, wish it were more like *Walden,* exploring the history of the social world as well as the natural history of the woods. Before deciding on the merits of this criticism, we need to consider what this argument ignores.

1

In the early seventies, during the time Dillard wrote her "mystical excursion into the natural world,"[2] she was living with many close neighbors near Roanoke, Virginia, in the immediate vicinity of Hollins College. She lacked the perspective to engage in a social discussion in the manner of Thoreau: "This is, mind you, suburbia," she says in a later essay: "It is a five-minute walk in three directions to rows of houses.... There's a 55 mph highway at one end of [Hollins] pond, and a nesting pair of wood ducks at the other. Under every bush is a muskrat hole or a beer can."[3] Her overriding concern is how to convey the idea of the wilderness, how to recover the aura of the frontier. Dillard's perception of the steers that graze on a nearby pasture illustrates the difficulty she has of even locating the boundary between nature and society. They are a constant reminder of the encroachment of the social order.

> They are all bred beef: beef heart, beef hide, beef hocks. They're a human product like rayon. They're like a field of shoes. They have cast-iron shanks and tongues like foam insoles. You can't see through to their brains as you can with other animals; they have beef fat behind their eyes, beef stew.[4]

There are other emblems of society that turn up in the woods. On a high hill, where Dillard watches "an extended flock of starlings" each dusk, someone had unloaded a pile of burnt books, "cloth- and leather-bound novels, a complete, charred set of encyclopedias decades old, and old watercolor-illustrated children's books" (*TC,* 39). On the quarry path she finds a discarded aquarium. Why someone hauled it so far into the woods to throw away is a mystery to her. But taken together these cultural artifacts suggest a renouncement of society and a commitment to the exploration of nature: "I could plant a terrarium here. . . . I could transfer the two square feet of forest floor *under* the glass to *above* the glass, framing it . . . and saying to passers-by look! look! here is two square feet of the world" (*TC,* 262). She cherishes the same dream as Thoreau: "If America was found and lost again once, as most of us believe, then why not twice?"[5] Only by relinquishing her hold on society can she imaginatively revive the promise of the new world.[6]

This is not to suggest that the criticism of *Tinker Creek* is unfounded. Dillard sometimes diverts our attention from society when a greater discussion seems warranted by her own narrative. "Sitting under a bankside sycamore" one day, Dillard contemplates what it means to live in the present. Before she can direct her attention to the natural world around her, she must first acknowledge the powerful undercurrents of the unconscious. She is daydreaming, fantasying: "I am in Persia, trying to order a watermelon in German." Continuing in this vein, she sees "the tennis courts on Fifth Avenue in Pittsburg, an equestrian statue in a Washington park, a basement dress shop in New York City—scenes that I thought meant nothing to me" (*TC,* 93). This may be. But is it not incumbent upon a writer who claims that nature "is my city, my culture, and all the world I need" (*TC,* 213) to explain the persistence of such images which inhibit her ability to concentrate on the natural phenomena at the creek?[7]

Yet if the weakness of *Tinker Creek* "resides in what it fails to consider,"[8] it is curious that the few passages which do deal with society are unsatisfying. Dillard is criticized for not doing more of what she does least convincingly. We need to consider what we are asking of her. "If the endings of our pastoral fables are generally unsatisfactory," says Leo Marx,

> if they seem to place the protagonist in equivocal, self-contradictory postures, it is largely because of the seemingly insoluble dilemma in which he has been put. How can he carry back into our complex social life the renewed sense of possibility and coherence that the pastoral interlude has given him? None of our writers has been able to find a satisfactory answer to this question.[9]

While it is true that Thoreau devotes a considerable amount of attention to society in *Walden,* he did not attempt to find a middle ground between it and his experiment in the woods. He was not any more than Dillard a social engineer. "He wished for a government so well administered," says William Bronk, "that private men need never hear about it."[10] "In one half-hour," Thoreau boasted, "I can walk off to some portion of the earth's surface where a man does not stand from one year's end to another, and there, consequently, politics are not, for they are but as the cigar-smoke of a man."[11] When he saw the extent to which the government of Massachusetts was willing to impinge upon the freedoms of its citizens to preserve the institution of slavery, he felt obliged to lash out at the state, but "he was radically inconsistent in his approaches to reform, approaches that included passive resistance, violence, and indifference."[12]

Such diverse approaches to social change were also manifest in the sixties, when Thoreau, not surprisingly, reached his greatest popularity, and when Dillard, in the midst of Watergate, wrote a book against the currents of the time.

> The kind of art I write is shockingly uncommitted—appallingly isolated from political, social, and economic affairs. There are lots

of us here. Everybody is writing about political and social concerns; I don't. I'm not doing any harm.[13]

The note of defensiveness is clear. She is up against a strong cultural trend. As Michael Meyer explains it:

> Every age is a political one, but ours is consciously so. Today we tend to make sense of ourselves and the world from a political perspective. Whereas the Puritans saw the hand of Providence or the claw of Satan informing every natural event, our own age has opted for politics as the most powerful means of explaining phenomena.[14]

It is not surprising that Dillard is criticized for ignoring the towering political institutions of the day.[15]

2

The appearance of other people in *Walden* and *Pilgrim at Tinker Creek* is another measure of the writer's attitudes toward society. While Thoreau's friends and acquaintances make up a large part of his narrative, Dillard's do not, despite the fact that she mentions many of their names, for example, the Whites, the Garretts, Maren and Sandy, Thomas McGonigle, Rosanne Coggeshall, Matt Spireng, and "Sally Moore, the young daughter of friends." These are merely references, however. The social interchange is usually offstage. Her description of playing baseball tells us more.

> My mind wanders. Second base is a Broadway, a Hollywood and Vine; but oh, if I'm out in right field they can kiss me goodbye.... I have no idea how many outs there are; I luck through the left-handers, staring at rainbows. The field looks to me as it must look to Wes Hillman up in the biplane: everyone is running, and I can't hear a sound. The players look so thin on the green, and the shadows so long, and the ball a mystic thing, pale to invisibility. (*TC*, 107–8)

Her playing parallels the movement of her book. She moves so far away from society that she sees it only as a world of shadows and symbols. She is a displaced romantic, out of touch with the fashion of the time.

> I have often noticed that these things, which obsess me, neither bother nor impress other people even slightly. I am horribly apt to approach some innocent at a gathering and, like the ancient mariner, fix him with a wild, glitt'ring eye and say, "Do you know that in the head of the caterpillar of the ordinary goat moth there are two hundred twenty-eight separate muscles?" The poor wretch flees. I am not making chatter; I mean to change his life. (*TC*, 132)

We are reminded of Thoreau:

> Sometimes when, in a conversation or a lecture, I have been grasping at, or even standing and reclining upon, the serene and everlasting truths that underlie and support our vacillating life, I have seen my auditors standing on their *terra firma*, the quaking earth ... watching my motions as if they were the antics of a ropedancer or mountebank pretending to walk on air.[16]

Dillard knows the danger, we may assume, of a too-close association with the legends surrounding Thoreau. She does not want to be perceived as a hermit, so she speaks of people in passing, assuring us she is part of the community. But by naming her goldfish Ellery Channing, a teasing reference to "the most intimate and most lasting friendship of Thoreau's life,"[17] she announces firmly that her personal life will remain private. At the same time she commits herself to a metaphorical rendering of her life in society through her relationships and encounters with the animal world. Although the goldfish only cost her twenty-five cents at a local pet store, she realizes that this simple transaction—"I handed the man a quarter, and he handed me a knotted plastic bag bouncing with water in which a green plant floated and the goldfish swam"—involves the intricate mystery of a living thing. "This Ellery" has "a coiled gut, a spine radiating fine bones, and a brain." He also has a heart. Dillard recalls how years ago she looked through a powerful microscope and saw the individual red blood cells of a goldfish pulsating through a section of its transparent tail.

> I've never forgotten the sight of those cells; ... I think of it lying in bed at night, imagining that if I concentrate enough I might be able to feel in my fingers' capillaries the small knockings and flow

of those circular dots, like a string of beads drawn through my hand. (*TC,* 125)

This rosary of cells, a symbol of the sanctity of life, is a common bond between Dillard and the fish, between animal life and human life in general, and between Dillard and other people. The symbolic richness of the goldfish is expanded by Dillard's later discussion of the spiritual significance of the fish in Tinker Creek, and especially by the affinity between fish and the early Christian church.

Ellery Channing is Dillard's most powerful cultural symbol, representing her detachment from society as well as her acknowledgment of the common bond of all living things. It is the detachment, however, which carries the greatest weight. Her relationship to the social world in *Tinker Creek,* and what may be read as her underlying attitude toward it, is ambiguous and largely indifferent. She does not succeed in encompassing within her vision any but the most fragmentary consequences for society at large. The Pilgrims' errand into the wilderness was a community's joint venture; the transcendentalists' return to nature was a limited fellowship of the spirit; Dillard's exploration of the Virginia woodlands is the solitary search of the soul.

Dillard's social isolation is seen most clearly in "Flood," where the overflowing creek and the coming together of the community to combat it paradoxically suggest even greater detachment than those chapters which do not mention society at all. The narrative is a flashback, brought to mind by the first pounding rain of summer. It signifies a switch in the emphasis of *Tinker Creek* from a celebration of nature to an accounting of its grimmer aspects. The intricacies of the year's new life are now caught up in a headlong rush toward destruction.

> That morning I'm standing at my kitchen window. Tinker Creek is out of its four-foot banks, way out, and it's still coming. The high creek doesn't look like our creek. Our creek splashes transparently over a jumble of rocks; the high creek obliterates everything in a flat opacity. It looks like somebody else's creek that has usurped or eaten our creek and is roving frantically to escape, big and ugly, like a blacksnake caught in a kitchen drawer. (*TC,* 149)

It is the incursion of nature into society that necessitates Dillard's sudden recognition of Tinker Creek as a part of a larger world. It is now "our creek," because "like a blacksnake caught in a kitchen drawer," it presents a social problem. But although the narrative suggests a meaningful coming together of neighbors to observe and combat the ravages of nature, the overall impression is one of mild coexistence. The people come together quickly and disperse even faster. They stand like spectators or so many vagrants poking at the remains of some ruin.

"Neighbors who have barely seen each other all winter," Dillard says, "are there, shaking their heads" (*TC,* 150). One wonders about the social isolation during the fairer months of spring. Dillard stands on a bridge over the torrent of water only inches below. She sees the crumbling ruins of a civilization hurtled by: dolls, split wood and kindling, bottles, rakes and garden gloves, railroad ties, lattice fencing, and a wooden picket gate. "There are so many white plastic gallon milk jugs," she says, "that when the flood ultimately recedes, they are left on the grassy banks looking from a distance like a flock of white geese." Only this impressionistic rendering of the aftermath of the flood can recall the pastoral landscape. One joint activity proves sadly ineffectual. A road crew is unsuccessful in attempting to free a tree trunk wedged solidly into a bridge's railing. Using a long-handled ax, they barely make a dent. "It's a job for power tools," Dillard concludes (*TC,* 151–52).

The narrator participates in helping to roll away some metal drums so a truck bearing a much-needed water pump can make it over a bridge. The crowd cheers in appreciation. This is the closest we come to any dialogue in the chapter, except for the reported fact that men lower their folded newspapers when speaking "and squint politely into the rain." This mute fellowship is nowhere more emphatic than when Dillard, who is walking on a narrow brick wall which extends into the flood, meets a young man going the opposite way: "The wall is one brick wide; we can't pass. So we clasp hands and lean backwards over the turbulent water; our feet interlace like teeth on a zipper, we pull together, stand, and continue on our ways" (*TC,* 155). This "marks the rare appearance," Eudora Welty says, "momentary as it is, of another human being in her book, and the closest any human being comes into the presence of the author."[18] When the flood waters recede, Dillard's interest in the event, people and all, diminishes as well: "The water's going down ... and the danger is past. Some kids start doing tricks on a skateboard; I head home" (*TC,* 156).

The exacting limits of this social exchange set the narrator more apart than ever. The townspeople meet as strangers, cooperating out of a shared danger. Nature offers no bridge to human relationships, but merely accentuates the common struggle for survival. One metaphor, significant as free association, most clearly conveys Dillard's attitude toward the people of the town. Some of the women, she observes, "are carrying curious plastic umbrellas... they don't put up, but on; they don't get under, but in. They can see out dimly, like goldfish in bowls" (*TC,* 155).

Exactly one year later, on "June twenty-first, the solstice, midsummer's night, the longest daylight of the year," Dillard returns to the creek to investigate the effects of the first heavy rain of summer. "High water had touched my log, the log I sit on, and dumped a smooth slope of muck in its lee." The creek is opaque and lightless. The day has "an air of meanance."

> A knot of yellow, fleshy somethings had grown up by the log. They didn't seem to have either proper stems or proper flowers, but instead only blind, featureless growth, like etiolated potato sprouts in a root cellar. I tried to dig one up from the crumbly soil, but they all apparently grew from a single, well-rooted corm, so I let them go. (*TC,* 148)

This blind featureless growth is perhaps the emblem of a society which the narrator finds impossible to describe. It all seems to grow from a single, well-hidden source. She lets it alone.

3

Although Thoreau and Dillard characteristically turn away from society, they do not lose their social instinct. Both seek community in the woods. Thoreau makes a companion of a mouse, who comes out "regularly at lunch time" and picks up the crumbs at his feet.

> It probably had never seen a man before; and it soon became quite familiar, and would run over my shoes and up my clothes.... At length, as I leaned with my elbow on the bench one day, it ran up my clothes, and along my sleeve, and round and round the paper which held my dinner, while I kept the latter close, and dodged and played at bo-peep with it; and when at last I held still a piece

of cheese between my thumb and finger, it came and nibbled it, sitting in my hand.[19]

Thoreau also becomes familiar with the birds, "not by having imprisoned one, but having caged myself near them" (*W,* 85). A phoebe soon builds in his shed, "and a robin for protection in a pine which grew against the house" (*W,* 226). Even the shy partridge leads her brood past his windows. The innocent expression of these young birds, he notes, suggests "not merely the purity of infancy, but a wisdom clarified by experience" (*W,* 227). This is a wisdom he seldom finds in society. It is as rare as the "principle" that dictates the battle of the ants: "There is not the fight recorded in Concord history, at least, if in the history of America, that will bear a moment's comparison with this, whether for the numbers engaged in it, or for the patriotism and heroism displayed" (*W,* 230). This may be why he dates the battle "in the Presidency of Polk, five years before the passage of Webster's Fugitive-Slave Bill" (*W,* 232). Were men so principled, his juxtaposition implies, the law would never have passed.

In these neighboring woods are also an occasional dog, a "domestic" cat, and a loon he plays "checkers" with on the pond. During winter, a squirrel wakes him at dawn, scrambling "over the roof and up and down the sides of the house, as if sent out of the woods for this purpose" (*W,* 273). He, in turn, feeds the squirrels, as well as the rabbits, blue jays, chickadees, and titmice. The latter become "so familiar that at length one alighted on an armful of wood which I was carrying in, and pecked at the sticks without fear." The squirrels also become "familiar, and occasionally stepped upon my shoe, when that was the nearest way" (*W,* 276). A hooting owl resides nearby, and although Thoreau never sees it, its "how der do" seems friendly enough, even if, at other times, its voice reminds him of the "dying moans of a human being" (*W,* 125). It is perhaps the cat-owl who most emphatically defines the community. As if hooting from the very citadel of the woods, it bids a traveling flock of noisy geese to leave the neighborhood in peace. This remarkable discord, Thoreau observes, had within it the "elements of a concord such as these plains never saw nor heard" (*W,* 272). The confederacy of animals in Thoreau's woods is stronger and more united than the troubled republic without.

Unlike the animals that live with Thoreau or treat him as a neighbor, the animals in Dillard's woods tend to flee from her. She is as alien in the woods as she is in society: "The creatures I seek have several senses and free will; it becomes apparent that they do not want to be seen" (*TC,* 184). She identifies instead with their wariness, their essential vulnerability.

> I am a frayed and nibbed survivor in a fallen world, and I am getting along. I am aging and eaten and have done my share of eating too. I am not washed and beautiful, in control of a shining world in which everything fits, but instead am wandering awed about on a splintered wreck I've come to care for, . . . whose bloodied and scarred creatures are my dearest companions. (*TC,* 242)

She had not always felt so. As Thoreau in his early life once hunted, Dillard used to kill and collect bugs.

> I used to kill insects with carbon tetrachloride—cleaning fluid vapor—and pin them in cigar boxes, labeled, in neat rows. . . . I quit when one day I opened a cigar-box lid and saw a carrion beetle, staked down high between its wing covers, trying to crawl, swimming on its pin. (*TC,* 52)

Now she is not surprised when animals flee: they have learned the lesson of the jungle. When she walks through a field teeming with grasshoppers, not one of them senses her peaceful intent: "To them I was just so much trouble, a horde of commotion, like any rolling stone" (*TC,* 210). A green heron, trying to feed along the creek, watches Dillard instead, "as if I might shoot it, or steal its minnows for my own supper" (*TC,* 187). These animals are her "companions at life." She does not dismiss them, even at home.

> I allow the spiders the run of the house. I figure that any predator that hopes to make a living on whatever smaller creatures might blunder into a four-inch square bit of space in the corner of the bathroom where the tub meets the floor, needs every bit of my support. (*TC,* 50)

These life and death entanglements in *Tinker Creek* function as a parallel to the sinister elements of society, illustrating the dimensons of the social world which have prompted her retreat.

Yet implicit in Dillard's narrative is the hope that by enacting the predatory pattern of the animal world, by stalking animals herself, she might reverse the estrangement of its creatures by showing them that she means no harm. She sets herself no easy task in pursuing the muskrat, who is so cautious, some believe, that he is impossible to observe. "They show me by their very wariness what a prize it is simply to open my eyes and behold" (*TC,* 192). She has not been the first to see this. He is the most celebrated animal in Thoreau's *Journal,* symbolizing a "stoic and unassuming courage."[20] "While I am looking at him," Thoreau says, "I am thinking of what he is thinking of me. He is a different sort of man, that is all."[21] The most memorable entries discuss the muskrat's willingness to gnaw off its own leg to free itself from a trap. Some have reportedly done so more than once, finally unable to escape.

> Shall we not have sympathy with the muskrat which gnaws its third leg off, not as pitying its sufferings, but through our kindred mortality, appreciating its majestic pains and heroic virtue? For whom are psalms sung and mass said, if not for such worthies as these? ... Prayer and praise fitly follow such exploits. ... Even as the worthies of mankind are said to recommend human life by having lived it, so I can not spare the example of the muskrat.[22]

Nor can Dillard, whose devotion is rewarded tenfold. One evening as she sits on a pedestrian bridge traversing one of Tinker Creek's small feeder streams, a muskrat emerges from his den, swims under the bridge, and commences feeding on the bank nearby. Soon, sliding back into the water and crossing under the bridge, he climbs up on a submerged rock and finishes his weed. Dillard is not four feet away. Returning to the same place on the bank, he continues his meal, until, gathering a huge mouthful of grass and clover, he dives into the water again. This time, however, he does not return to the rock. He appears on the bank at her side. "I could have touched him with the palm of my hand." Dillard watches transfixed while for several moments he rummages in the grass before returning to the water and vanishing into his den.

Though exhilarated by the experience, Dillard sometimes desires more. If we are all victims, can we not meet a higher level of trust and understanding? Can we not celebrate our sacred ordinary noons? Haunted daily by the memory of the frog which had been eaten by the giant water bug (*TC,* 5–6), Dillard ventures to a nearby duck pond once more to attempt a playful fellowship with frogs. Except for one ominous reminder of her previous experience, her celebratory communion is observed.

> Tonight I walked around the pond scaring frogs; a couple of them jumped off, going, in effect, eek, and most grunted, and the pond was still. But one big frog, bright green like a poster-paint frog, didn't jump, so I waved my arm and stamped to scare it, and it jumped suddenly, and I jumped, and then everything in the pond jumped, and I laughed and laughed. (*TC,* 118)

4

The obscure portrait of society in *Tinker Creek* brings to mind the question of how different the book would be had it attempted to take into account the surrounding community. An essay by Dillard published the following year shows what the book might have gained and lost by a more overt social discussion.[23] The persona is the same, narrating a late afternoon walk across the creek and up a "hill far away" to a place she had never been before. Although she can still see the creek from the top of the hill, she fixes her attention on a young boy. He is standing by a toolshed in the back property of a large estate. He is alternately pretending to write on the wall with a stone and fooling with two frantically barking dogs in a pen. He does not connect the dogs' behavior with Dillard's presence. When he finally sees her, he comes over to the barbed-wire fence to converse. He is an articulate but insipid "bourgeois gentilhomme" who looks "like a nineteenth-century cartoon of an Earnest Child." He rouses little interest in Dillard: "What was I doing chatting with a little kid? Wasn't there something I should be reading?" (It is an attitude which partly explains the absence of silmilar occurrences in *Tinker Creek*.) What happens next, however, not only renews her interest in the child but places the relationship in a broader social context.

> He paused. He looked miserably at his shoetops, and I looked at his brown corduroy cap. Suddenly the cap lifted, and the little face said in a rush, "Do you know the Lord as your personal savior?" "Not only that," I said, "I know your mother." It all came together. She had asked me the same question. Until then I had not connected

this land, these horses, and this little boy with the woman in the big house at the top of the hill, the house I'd approached from the other direction, to ask permission to walk the land. There had been a very long driveway from the highway on the other side of the hill. The driveway made a circle in front of the house, and in the circle stood an eight-foot aluminum cross with a sign underneath it reading **CHRIST THE LORD IS OUR SALVATION**. . . .

The woman was very nervous. She was dark, pretty, hard, with the same trembling lashes as the boy. . . . My explanation of myself confused her, but she gave permission. . . . She was worried about something else. She worked her hands. I waited on the other side of the screen door until she came out with it.

"Do you know the Lord as your personal savior?"

My heart went out to her. No wonder she had been so nervous. She must have to ask this of everyone, absolutely everyone she meets. That is Christian witness. It makes sense, given its premises. I wanted to make her as happy as possible, reward her courage, and run.

The boy is happy that Dillard knows his mother, but he seems "tired, old even, weary with longings, solemn." He never plays at the creek "because he might be down there, and father might come home not knowing he was there, and let all the horses out, and the horses would trample him." There are also snakes. This cautiousness prompts Dillard to encourage him to play there; she tells him that she does so herself. Yet it also makes her question why this should be important: "What do I do there alone that he'd want to do? What do I do there at all? Why would anyone in his right mind play at the creek?" She has never so clearly faced the question of its practical social value. That she does not attempt to answer the queston in the essay is not an evasion. She does not know. She cannot translate "the renewed sense of possibility and coherence" that the pastoral interlude has given her.

Although merely implied by her questions, the problem seems to lie in the contrast between the fundamentalist religion of the family and the spiritual-aesthetic faith of *Tinker Creek*. What, if anything, does her literary refashioning of nature have to do with that eight-foot aluminum cross? What is art compared to personal salvation? The delicate framework of Dillard's belief seems unable to sustain the weight of practical inquiry.

The spiritual-aesthetic experience which Thoreau and Dillard cultivate in the woods is by necessity a solitary one. The Puritans did not celebrate mass but awaited the direct indwelling of God. There was not one religious experience but many private ones. It is not surprising that John Field catches no fish when he finally joins Thoreau or that Dillard is unsuccessful in showing muskrats to other people. Whenever two or more gather in the woods, the likely outcome is a manifestation of society, not religion. Annie Dillard goes into the woods to claim her spiritual heritage. Like a prophet she travels alone.

NOTES

1. "Attractions and Dangers of Nostalgia," review of *Pilgrim at Tinker Creek, Virginia Quarterly Review* (Autumn 1974): 640. For similar, but less severe, responses, see the reviews by Eva Hoffman and Eudora Welty ("Meditation on Seeing") listed in notes 8 and 18.

2. Subtitle of the paperback edition of *Pilgrim at Tinker Creek* (New York, 1975).

3. "Living Like Weasels," in *Teaching a Stone to Talk* (New York, 1982), 12–13.

4. *Pilgrim at Tinker Creek* (New York, 1974), 4: subsequent page references will be cited in the text as *TC*.

5. Henry David Thoreau, *Cape Cod* (Boston, 1865), 231.

6. In an interview with Karla M. Hammond ("Drawing the Curtains: An Interview with Annie Dillard," *Bennington Review* 10 [April 1981], 34), Dillard claims that the first person, as she uses it, "is merely a narrative device—a kind of floating eyeball, a unifying voice." Although I believe that this exaggerates how the narrative voice functions in *Tinker Creek,* it might be used as another defense against those critics who feel that the narrator should be more involved with the community.

7. See *TC* 48–49, 81, and 224–25, where a greater discussion of society also seems warranted.

8. Eva Hoffman, "Solitude," review of *Pilgrim at Tinker Creek, Commentary,* October 1974.

9. "Pastoral Ideals and City Troubles," *Journal of General Education* 20 (January 1969): 266.

10. *The Brothers in Elysium: Ideas of Friendship and Society in the United States* (New Rochelle, 1980), 64.

11. "Walking," *The Natural History Essays* (Salt Lake City, 1980), 101.

12. Walter Harding and Michael Meyer, *The New Thoreau Handbook* (New York, 1960), 137.

13. Quoted in Philip Yancey, "A Face Aflame: An Interview with Annie Dillard," *Christianity Today,* 5 May 1978, 16.

14. *Several More Lives to Live:* Thoreau's Political Reputation in America (Westport, CT, 1977), 4–5.

15. It is ironic that ten years later, in an essay for the *Esquire* series "Why I Live Where I Live" (March 1984, 90–92), Dillard writes what may be regarded as the "missing" chapter of her book. The tone is unmistakably Thoreauvian even if her conclusions are not: "I came to Connecticut because, in the course of my wanderings, it was time to come back east—back to that hardwood forest where the multiple trees and soft plants have their distinctive seasons and their places in sun and shade.... So why, if I came for the forest, don't I live in the forest?... I distrust the forest, or any wilderness, as a place to live.... I am a social animal, alive in a gang, like a walrus, or a howler monkey, or a bee.... So I live, and have almost always lived, in association with a college campus. I need lots of time, a big house with room for lots of desks, and a batch of people to befriend—people who are also working among bits of paper all day."

16. *The Journal of Henry D. Thoreau,* ed. Bradford Torrey and Francis H. Allen, 14 vols. (Boston, 1906), 9: 237–38.

17. Walter Harding, *The Days of Henry Thoreau* (New York, 1982), 173.

18. "Meditation on Seeing," review of *Pilgrim at Tinker Creek, The New York Times Book Review,* 24 March 1974, 5.

19. *Walden,* ed. J. Lyndon Shanley (Princeton, 1971), 225–26; subsequent references will be cited in the text as *W.*

20. Donald M. Murray, "Thoreau and the Example of the Muskrat," *Thoreau Journal Quarterly* 10 (October 1978): 3.

21. *Journal,* 2:111.

22. *Journal,* 6:98–99.

23. "On a Hill Far Away," *Harper's,* October 1975, 22–25; reprinted in her *Teaching a Stone to Talk,* 77–83. Quotations are taken from the slightly revised version in *Teaching a Stone to Talk.*

Joseph Wood Krutch
1893–1970

> In the animal kingdom, monogamy, polygamy, polyandry and promiscuity are only trivial variations. Nature makes hermaphrodites, as well as Tiresiases who are alternately of one sex and then of the other; also hordes of neuters among the bees and the ants. She causes some males to attach themselves permanently to their females and teaches others how to accomplish impregnation without ever touching them. Some embrace for hours; some, like Onan, scatter their seed. . . . To her children nature seems to have said, "Copulate you must. But beyond that there is no rule. Do it in whatever way and with whatever emotional concomitants you choose. That you should do it somehow or other is all that I ask."
>
> —Joseph Wood Krutch, *The Voice of the Desert*

Joseph Wood Krutch was born in Knoxville, Tennessee, in 1893, the youngest of three brothers. He graduated from the University of Tennessee in 1915 and moved to New York to continue his study of literature, earning a Ph.D. in English from Columbia University. He soon became a drama critic for *The Nation*, a position he kept for many years. He taught English and drama on a part-time basis at a variety of colleges, eventually receiving an appointment as professor of English at Columbia. Between these two spheres of interest, theater criticism and nondramatic literature, Krutch developed an impressive publishing record of at least fifteen books, including major critical biographies of Samuel Johnson, Edgar Allen Poe, and Henry David Thoreau.

His Thoreau biography and a visit to the Southwest in 1939 were important stepping stones toward Krutch's second career, as a nature writer. The Krutches' first summer holiday trip to Arizona was followed by annual pilgrimages to the southwest until the war and gas rationing prevented them. After the war, the Krutches spent a sabbatical year in Tucson in 1950, during which Krutch wrote *The Desert Year* (1952). He retired from Columbia soon afterward and moved to a permanent home in Tucson. His bibliography includes at least ten books of nature writing, including the following major works: *The Desert Year* (1952), which records Krutch's growing love affair with and understanding of the flora and fauna of the desert; *The Voice of the Desert* (1955), perhaps his most insightful and profound study of desert life; *Grand Canyon* (1958), which took readers on a personal tour of the geology, history, botany, and zoology of the big canyon; and *Baja California and the Geography of Hope* (1967), which reveals the remote beauty of this little-known region.

Although Krutch was a champion of evolutionary theory, he believed in an evolution of abilities and qualities quite beyond a strictly mechanistic theory of survival of the fittest. He became particularly interested in those desert plants and animals that had made radical adaptations to survive in the hostile desert environments. Although he left his academic career and urban orientation in despair over the loss of traditional values in America and in the humanist tradition, he discovered in the desert other, more fundamental, values concerning the nature of life in a world not made by man. Joseph Wood Krutch's desert career produced major contributions to American nature writing. Because of his popularity, he was able to bring conservation ideas to a wide readership. In his capacity as a staunch advocate for national parks and for wilderness protection, he declared, "the wilderness and the idea of wilderness is one of the permanent homes of the human spirit."

"Mr. Krutch"

Edward Abbey

Like most people who knew him, I first met Joseph Wood Krutch in the pages of his books. During my student days I read *The Modern Temper* and was pleased by its austere pessimism. Then I read his studies of Poe, Samuel Johnson, and Thoreau, and later, at approximately the same time that I was beginning to think about the subject myself, I read his justly popular books concerning the Southwest—*The Desert Year* and *The Voice of the Desert*. In the voice of Mr. Krutch I found the clearest definition in contemporary literature of what many of us have felt, emotionally, intellectually, instinctively, to be the central meaning of the word *civilization*.

Civilization, if it means anything and if it is ever to exist, must mean a form of human society in which the primary values are openness, diversity, tolerance, personal liberty, reason. It appears doubtful that such a society has existed in the past and at present more doubtful that it will come to be in the near future—that is, within the next century or two (until some debris has been cleared away). Nevertheless, civilization as here defined seems to me the one clear purpose implied in the martyrdom of prophets, the wisdom of seers and poets and thinkers, the suffering and torture of common people, that have characterized human history for the past five hundred years. If there is such a thing as human evolution (and I suspect there is) then the slow, painful effort toward a free community of men and women, with a full flowering of the individual personality, must be the ideal—always opposed but never wholly suppressed—which has inspired our long travail. If there is no such goal, then human history is indeed, as some have called it, nothing but a nightmare.

Among a thousand others one could name (Mann, Tolstoy, Kropotkin,

Camus, Russell, William O. Douglas, Lewis Mumford, Aldous and Julian Huxley, for example), it seemed and still seems to me that Joseph Wood Krutch, in his modest but steadfast way, was one of the soldiers on the side of what we may properly term progress—not as the politicians and technocrats have profaned the word, but progress in its larger sense, as the tortuous advance toward the idea of civilization. Mr. Krutch's contribution to this campaign has been his communication of the discovery that the natural world must be treated as an equal partner. A world entirely conquered by technology, entirely dominated by industrial processes, entirely occupied by man and machine, would be a world unfit to live in—perhaps impossible to live in. The technosphere of R. B. Fuller, the global village of McLuhan, the arcologium of Soleri, the noosphere of de Chardin, the planetary suburb of Herman Kahn, and similar concepts of other such visionaries (most of them already archaic) would be the ultimate trap for humankind, threatening us with the evolutionary dead end of the social insects—the formicary of the ants, the termite commune.

Mr. Krutch did not deal, in his "nature" books, with these global prospects, but confined his studies to the effects of the conflict in his own backyard, that is, the deserts of Arizona, Sonora, and Baja California, and the Grand Canyon of the Colorado. But as in any careful work, the results of regional investigations have planetary significance.

What I most admired in Krutch's work was not what he said—others said it too—but the way in which he said it. Many Americans and a few Europeans, influenced by the long Western tradition beginning with the heretical Saint Francis and carried on by Thoreau, were coming to share the same belief in the right of the nonhuman world to exist—to exist not only for the pleasure and health and instruction of humankind but for its own sake. Not only the pretty birds but also the predators and reptiles, the ugly and unloved, the organic and inorganic—all belong here with us, on the same small planet. In some sense of the word a boulder on a mountainside has a *right* to be left undisturbed, to *enjoy* its own metamorphoses in its own way at its own nonhuman pace of change. Mr. Krutch came closer than any other contemporary American writer to supplying for such supposedly nonsensical notions the necessary supporting structure of rational thought.

Rational thought. Calm, reasonable, gentle persuasion. It was this quality of moderation in his writing that most impressed me, for my own inclinations always tended toward the opposite, toward the impatient, the radical, the violent. A book, I often thought, was best employed as a kind of paper club to beat people over the head with, to pound them into agreement or insensibility. Fully aware of my self-defeating tendencies as a propagandist, I suspected it would be both interesting and useful to meet Mr. Krutch in person, face-to-face. Since we were both living in Tuscon at the time I discovered his work, I soon found a pretext for paying him a visit.

The editor of an ephemeral magazine called *Sage* (referring not to sagacity but to a certain shrub popular in the American West) agreed to sponsor me as an interviewer of our most distinguished Southwestern author. I found Krutch's name and address listed in the telephone book and gave him a ring. Although he had never heard either of me or of *Sage,* he agreed readily to an interview. A few days later I knocked on his door. This was sometime in the winter of 1967–68; what follows is an account of the last formal interview that Krutch ever granted. At the time I did not know he was seriously ailing; he died not long afterward. Despite the triviality of my questions, our conversation may be of some interest to students and admirers of Krutch and his books.

Mr. Krutch greeted me at the door, invited me in. He looked like a professor—thin, gray, a trifle abstracted. He looked older than I expected. We sat down in the scholar's traditional book-lined study.

As befitted his manner, appearance, and prose style, his speech was formal, almost pedantic. In other words he spoke in complete sentences and unified paragraphs, each thought leading logically into the next, like a man accustomed to lecturing, or like a man accustomed to thinking, an odd sort of old bird. Tape recorder in motion, I began with the amateur interviewer's dumb initial question: "Mr. Krutch, what do you think of interviews?"

He said he was willing to risk it. Having expected a much more lengthy reply, I found myself at a loss for the next question. Like many writers, I am a man of few words. Krutch meanwhile studied me gravely, waiting in silence, hands clasped on his stomach. I groped around for a while in the catacombs of my mind. (Have you ever been awakened from a good sleep by the sound of a crash somewhere deep inside the skull—that startling collapse, as of an eroded mudbank, when the whole tiers and galleries of damaged brain cells give way and topple down the talus slopes of the cerebellum?) I ventured this: "As a student of man and society, do you see much hope for our Southwestern cities? Tuscon, for example?"

"A few years ago," he replied, after a pause, "I was dragooned into giving a speech at a luncheon of the Rotary Club here. The assigned subject was 'Tuscon as a Place for the Writer,' and I said that Tuscon was still pretty good but getting less good every year. Whenever I see one of the billboards with the slogan 'Help Tuscon Grow,' I think, 'God forbid.' And so I suggested to the Rotary Club that the slogan be changed to 'Keep Tuscon Small.' They didn't follow my advice."

"Have you lived in other cities of the Southwest?"

"No, but I've seen them all."

"How do you like Las Vegas?"

"Las Vegas has a beautiful setting and the climate is very good. It might not be a bad place to live in. I don't suppose you have to see more of the Strip than you want to."

"Most people who do live there claim that, away from the Strip and Fremont Street, Las Vegas is much like any other American city."

"I suppose that's true, though I'm not sure that it should be considered complimentary. But the Strip for an occasional night is all right—if you like that kind of thing. You certainly won't find it held in such high esteem anywhere else."

"Have you been to the Grand Canyon lately?"

"I haven't been there in two or three years. Of course I got quite worked up about the threat of the dams. And I'm afraid that issue is not permanently disposed of yet. Many people here in Tuscon, if they have connections with the business and political community, think it scandalous to oppose the dams. For example, the editor of the *Tuscon Star* reviewed a book of mine a couple of weeks ago. He praised the book highly for most of the review, then condemned it in his last few paragraphs because of my opposition to the proposed dams."

"I saw that review. I think he was mostly upset by your criticism of hunting—hunting for sport, I mean. He seemed to think you were being unfair to sportsmen."

"I didn't know the piece would offend sportsmen. I was surprised by his indignation."

"Ever done any hunting yourself?"

"No. I hope I'm not fanatical on the subject, but I don't like the idea of killing for fun."

"Haven't the sportsmen usually been good allies of the conservationists?"

"Not always. Now up in the Kaibab Forest the hunters wanted an open season on the Kaibab squirrel—a rare species. And they would have got it, too, if the Forest Service hadn't opposed them. The sportsmen will say, in the case of deer, that you've got to thin out the population anyway. And then I say, even so, I object to hunters doing it—not so much for the sake of the deer as for the sake of the hunters. You see, I have this private, subjective feeling that killing things for the sake of sport is wrong. I think hunting is bad for hunters because killing for pleasure tends to brutalize those who do it. If it's a matter of thinning out excess populations, then by the hunter's logic we might as well start hunting each other."

Passing up my chance to defend sport hunting (I was once a sportsman myself—before I grew up), I suggested another topic: "Have you seen our newest national park—Canyonlands in Utah?"

"I've been on the edges of it."

"Have you heard about the proposals for the development of that park?"

"Only in a general way, but from what I've heard I hope they don't do it. On this question of development versus preservation, my belief is that some places should be open to maximum accessibility and others not. I think the idea that every national park should be developed in the same way is wrong."

"How about banning all cars from all parks?"

"I suppose I would approve of that idea, privately. Obviously it's not a practicable idea at present."

"Not politic?"

"Not politic. Too many people are accustomed to driving their cars through the parks. Take Yellowstone, for instance—we might as well let the auto tourists keep that one."

"A lost cause?"

"At present. But Canyonlands National Park is new, and as long as there are other easily accessible places in the area, like Arches National Park, with the same type of landscape, I can't see any need for new roads in Canyonlands. As I've said before, too many people use their automobiles not as a means to get to the parks but rather use the parks as a place to take their automobiles. What our national parks and forests really need are not more good roads but more bad roads."

"As a filtering device, you mean?"

"Exactly. There's nothing like a good bad dirt road to screen out the faintly interested and to invite in the genuinely interested. And it's perfectly fair and democratic, open to anyone willing to endure a little inconvenience and discomfort for the sake of getting away from the crowds."

Delving deeply, I managed to shovel up with mangled syntax a Serious Question: "Sir," I asked, "do you think there's any significant difference between man's relationship to the natural world in the desert and in other places, as in—well, say, New England?"

"I think that in both regions the opportunities to establish a relationship with the natural world are equally rich. And in both places there are many people who succeed in doing it. But I'm also afraid that in both there are too many people, as here in Tuscon, who seem to be completely unaware of the existence of a natural world. I mean the men preoccupied with industry, property sales, merchandising—they're completely indifferent to anything that lies beyond commercial possibilities. At least they behave as if they are. But the Southwest is one of the few places where nature *invites,* or almost *demands,* human contemplation. And it is one of the few places in our country where contemplation of anything but walls and bars and other human bodies is still possible."

Pause.

"Whatever we think of our Southwestern cities," I said, "at least they're easy to get out of. Unlike back East, where the cities are merging into one another."

"True. One can drive a few miles from Tucson, turn off on a dirt road passable to an ordinary car, and spend hours in the desert without seeing another soul."

"Do you feel any need to justify this preference of yours for privacy? For solitude?"

"None whatsoever."

"Do you think the desert is in any way intrinsically more interesting than other regions?"

"I suppose it's primarily a matter of temperament. For myself, I like the openness of the desert, the barrenness of desert mountains. I was brought up in Tennessee, and when I go back there and see those Appalachian mountains, beautiful though they are, I can't help feeling now that they're too heavily clothed in plant life; you can't see the basic forms. And also I suppose there's something dramatic about extremes— as in the desert and the jungle. In each, nature goes about as far as it can go in opposite directions. In one an extreme spareness, austerity; in the other an extreme of fecundity."

I said, "People used to regard the desert as ugly; now more and more seem to like it."

"Desert scenery appeals to modern taste—we now like spare, simple forms. Look at modern sculpture. And also, of course, the desert no longer seems dangerous. That makes a big difference. It must have been hard to appreciate the beauties of the desert when you were trying to get across it in a wagon train. In other words, for most people the wilderness does not become attractive until it is relatively safe. The eighteenth-century philosophers distinguished the 'awful' from the 'beautiful' by saying that the former suggested terror without being actually terrifying. Mountains, for instance, were 'horrid' until the eighteenth century made them merely 'awful.' Now that men are no longer terrified by mountains, we call them 'beautiful.'"

"Have you any favorite books or favorite writers on the subject of the desert? On nature in general?" (I can't believe I said that. But it's on the tape.)

"From the standpoint of factual information, I suppose Edmund Jaeger is as good as any on the Southwest. On the general subject of man's relation to the natural world, I prefer the biologists who take what I call an out-of-doors attitude as opposed to the laboratory outlook. For example, Loren Eiseley, a beautiful writer, and the American naturalist Marston Bates."

"How about George Gaylord Simpson? Julian Huxley?" I had special reasons for mentioning these two writers. I was once seated next to Mr. Huxley at a dinner. Since neither of us really wished to talk about evolution, we found a mutual interest in discussing the latest mutations from London: the Beatles and miniskirts. As for Dr. Simpson, the famed paleontologist, author of *Horses* and other books, whom my friend the storyteller William Eastlake considers the only authentic genius he has ever met, he too was living in Tuscon at this time. In company with Eastlake, I had recently spent an evening with Dr. Simpson and his wife at their home. On that occasion, after some loose talk on my part about the Peace Corps, which I dismissed as "a typical piece of American cultural insolence," I had the distinction of hearing the great George Gaylord Simpson call me, to my face, the stupidest young man he had ever met outside of Harvard. (To Simpson, anyone under seventy was a young man.) I don't recall finding an answer to that remark; mainly, I was pleased with myself at having riled the authentic genius so easily. It was one of the high points of my life. As we were leaving Dr. Simpson's house, rather shortly afterward, Eastlake informed me that Simpson's wife had been one of the prime organizers of President Kennedy's Peace Corps.

"Huxley and Simpson," Mr. Krutch was saying, "come more under the heading of formal science than of nature writing, but they're both among the best in their fields."

This was interesting. I had asked Dr. Simpson if he would like to meet Joseph Wood Krutch. He had said no. When I asked Mr. Krutch if he would like to meet Dr. Simpson, Mr. Krutch said yes. Is this one of the key differences between the formal scientist and a mere literary type—the wider tolerance and greater curiosity on the part of the latter?

I said, "Isn't Simpson's approach to biology a great deal more mechanistic than yours?"

"Yes, but nevertheless he doesn't go the whole way. I'd much rather read him than a full mechanist. And Julian Huxley makes what I think is a responsible concession to my point of view, in that he allows at least some role to the idea of purpose in the evolutionary process."

"Have you a name for your style of humanism?"

"I've come around, I think, to a form of philosophical monism, so I suppose you could call me a monistic humanist—or a humanistic monist. The terms don't mean much in themselves, but they do mean that I am still not convinced that the universe is merely a machine, in the limited sense of that word. That is, I don't think it means anything to speak of 'materialism,' for example, because the potentialities of matter have just recently begun to be realized. When matter can be transformed into energy, we don't really know what the true nature of matter is."

"Or mind? Or consciousness?"

"That too. So that if one says that 'vitalism' is dead, as a theory in biology, he'd better add that 'mechanism' is also dead. It seems to me that the world is all one thing, not two. Mind and matter are not opposites but aspects of something underlying both. Matter becomes less and less material, while mind is clearly one of the inherent potentialities of matter. But a lot of orthodox biologists are still unwilling to recognize this. They continue to think in nineteenth-century terms, assuming an absolute discontinuity between matter and consciousness, so that they're unable to explain how the latter can emerge from the former."

"There are some," I said, "who think the human mind is nothing but a miniaturized computer. Is that what you'd call laboratory thinking?"

"Yes. These are the same sort of technicians who used to compare the mind to a telephone system. But a telephone system never talks to itself. And neither does a computer. Some talk of the 'memory' of a computer, but there's no more justification for talking about the memory of a computer than the memory of a phonograph record. Both are mechanical means of bringing forth data or sounds previously placed there—storage systems, not memories."

"Couldn't the human memory be considered simply a storage system?"

"The human memory is a creative storage system. No computer has yet been able to create anything."

This was getting too deep for me; it was time to change the subject. "Speaking of creation, are you working on a book now?" (Every writer's favorite question.)

"I'm not working on a book, except insofar as I am continuing to write magazine essays which someday may be collected in book form. For the kind of subject I'm writing about, I probably get more readers in this way than I would in writing a book."

"Of your own books, which do you like best?"

"I think the one called *Twelve Seasons* was the most deeply felt thing I've done. But an early one, *The Modern Temper*—which, by the way, no longer suits my own temper—was the most popular of my books. No doubt because it was so pessimistic in outlook."

"Have you ever attempted to write a novel?"

"No. I'm afraid if I did, all my characters would tend to talk exactly like me. The nearest I've come to that sort of thing was in the imaginary dialogues between Johnson, Thoreau, and Shaw in my book *If You Don't Mind My Saying So.*"

"In which the characters quote from their own work."

"Yes. That way I made sure they wouldn't all sound like me."

"Are Thoreau and Samuel Johnson still among your intellectual heroes?"

"Oh yes."

"Aren't they nearly opposites, those two, intellectually and spiritually?"

"Yes, and like opposites they complement each other nicely. And they had this in common: both were uncompromising individualists."

"Do you keep up with current drama? Or with new fiction?"

"The last novelist I read with any great enthusiasm was Marcel Proust."

"What about James Joyce? Thomas Mann?"

"I've read both, and with interest. But neither quite suits me. I never could get through *Finnegans Wake*. Could you?"

"I've read the beginning and I've read the ending."

"Yes, I suppose that's the way most of us got through it."

"Do you still think Bernard Shaw will outlast them all?"

"Yes. I doubt if the others will be read a hundred years from now."

"How about some modern Americans—Hemingway? Faulkner?"

"Neither one of them have I read with any true satisfaction. In Hemingway I am repelled by that phony virility of his. I think it's a fake."

"He wrote good stories. Lots of good short stories."

"Oh yes. He really did have a gift for portraying character in a few words. When I read *A Moveable Feast* I was struck by the accuracy of his perceptions. I knew some of the people Hemingway writes about in that book, and even if he had not named them I would have recognized them easily, simply from his brief descriptions."

"Perhaps Hemingway was better at seeing the truth in others than—"

"Than in himself—that may be."

"And Faulkner?"

"I can't get through the jungle of his sentences."

"Because you're a desert lover."

"That might be the trouble."

I sensed that the good Mr. Krutch was getting bored with me—or tired. I couldn't blame him. I proceeded to wrap the thing up quickly. "You've written much on literature and drama, on other writers, on nature, on modern society, on contemporary science and thought. What would you say your chief interest is these days?"

"I'm still very much interested in nature and the spectacle of nature. But I suppose most of what I've been writing recently tends to center around the question of the reality of consciousness as opposed to a strictly mechanistic view of life, and around the opposition between traditional values and new technology . . . ."

The tape ends here. Mr. Krutch invited me to stay for lunch. I met his wife, Marcelle, a wise and witty woman; we talked of other things, including the Vietnam War. I was unable to get him to take a stand against what seemed to me then and seems to me now the most shameful episode in our nation's history since the time of legalized human slavery. Krutch, in early 1968, could not yet bring himself to believe ill of President Johnson, whom he much admired for his achievements in civil rights. Perhaps later he changed his mind; I do not know. He did not have much time left. I never saw him again.

In his "nature" books, Krutch explores in detail many of the themes mentioned in my depthless interview. He takes an overall view, beginning with protozoa and ending with the human—and with the song of a cardinal outside his window—finding in the latter the suggestion that joy, not the single struggle for survival alone, is the essence of life, both its origin and its object. One may quibble with some of his objections to Darwinian theory—the elegance of the idea of natural selection, for example, lies in the fact that it still appears, after a century and a half of scrutiny and attack, to serve as *sufficient* explanation for the method of evolution—without denying Krutch's thesis that there is far more in the character of living things, human and otherwise, than can be understood through statistical analysis, chemical reaction, or any kind of quantitive measurement, however refined. In his unwavering insistence, to the very end of his life, on the primacy of freedom, purpose, will, play, and joy, and on the kinship of the human with all forms of life, he defended those values which supply the élan vital of human history.

Joseph Wood Krutch was a humanist, one of the last of that endangered species. He believed in and he practiced the life of reason. He never submitted to any of the fads or ideologies or fanaticisms of the twentieth century. In his last completed book, *And Even If You Do* (following logically upon *If You Don't Mind My Saying So*), he quotes from a letter written by Marguerite Yourcenar, the historical novelist, to a friend.

> I tell myself that you are lucky to have lived during an epoch when the ideal of pleasure was still a civilizing idea (which today it no longer is). I am especially pleased that you have escaped the grip of the intellectuals of this century; that you have not been psychoanalyzed, and that you are not an existentialist nor occupied with motiveless acts; that you have on the contrary continued to accept the evidence of your understanding, your senses, and your common sense.

Madame Yourcenar's testimonial will serve as an apt description of the character of Joseph Wood Krutch himself. In this age of the myopic specialist and the cosmic expert, of the crazy, the extreme, and the violent, I can think of few higher tributes to offer to his memory.

Aldo Leopold
1886–1948

> We reached the old wolf in time to watch a fierce green fire dying in her eyes. I realized then, and have known ever since, that there was something new to me in those eyes—something known only to her and to the mountain. I was young then, and full of trigger-itch; I thought that because fewer wolves meant more deer, that no wolves would mean hunters' paradise. But after seeing the green fire die, I sensed that neither the wolf nor the mountain agreed with such a view.... The cowman who cleans his range of wolves does not realize that he is taking over the wolf's job of trimming the herd to fit the range. He has not learned to think like a mountain. Hence we have dustbowls, and rivers washing the future into the sea.
>
> —Aldo Leopold, "Thinking Like a Mountain,"
> in *A Sand County Almanac*

Aldo Leopold was born in Burlington, Iowa, in 1886. He attended Lawrenceville (New Jersey) preparatory school and then Yale University. Leopold received his B.S. in 1908, and in 1909 he earned a masters degree from the newly established School of Forestry. He began his career with the U.S. Forest Service and was soon assigned to Apache National Forest in Arizona Territory as an assistant forester. By 1917 Leopold had risen to the rank of assistant district forester. At this stage in the evolution of his career, Leopold's system for protecting valued game species involved near-extermination of their predators (wolves, coyotes, and mountain lions). He came to realize, however, the crucial role played by predators in keeping the prey species within the regional carrying capacity. From the 1920s on, his writing about wildlife management reflects his growing recognition of the need for ecosystem balance and long-term stability.

Leopold soon extended his emerging ecosystem theory, calling for a fundamental change in land-use designations in the national forest system. Along with the multiple use concept, on which the National Forest Service had been founded, he called for a separate and permanent wilderness designation. Despite widespread opposition, Leopold continued to argue his cause. His idea came to fruition in 1924, when the National Forest Service designated 574,000 acres in New Mexico as the Gila Wilderness Area. Leopold predicted that as the nation became increasingly urbanized, such undisturbed wilderness ecosystems would grow in value, for habitat protection, research, and controlled recreation.

Aldo Leopold left the National Forest Service in 1924 and served as associate director of the U.S. Forest Products Laboratory in Madison, Wisconsin, for the next four years. After several years of wildlife management consulting, he was appointed professor of wildlife management at the University of Wisconsin, a position he retained for the remainder of his career. During this phase of his work, Leopold made major contributions to soil conservation policy, wildlife management practices, and environmentally responsible agriculture. His book *Game Management* (1933) presented a revolutionary approach to his topic, based on the emerging science of systems ecology—an integration of population dynamics, habitat protection, and food chain research. At the philosophical center of his work is a shift from the prevailing view that nature is a resource commodity under human control to a belief that humans are a dependent part of a natural community with which they must learn to live in harmony. Leopold's best-known book, *A Sand County Almanac*, includes essays that fully developed his ecological philosophy, ethics, and aesthetics. There are many parallels between his *Almanac* and the nature writing of John Muir and Henry David Thoreau.

Leopold died in 1948 while fighting a forest fire. In the intervening years, his influence has continued to grow. Today he is considered an instrumental architect of the plans for an ecologically sustainable relationship between humans and the biosphere.

Anatomy of a Classic

John Tallmadge

Compared to other genres in the library of English literature, natural history writing fills only a modest shelf. Yet here too one can find genuine classics such as Gilbert White's *The Natural History of Selborne* (1789), Darwin's *Voyage of the Beagle* (1839), Thoreau's *Walden* (1854), and John Muir's *My First Summer in the Sierra* (1911). Like them, Leopold's *A Sand County Almanac* not only gratifies on first perusal but stimulates and refreshes even after many readings. Like all true classics, it only grows richer with time, and this is no mean accomplishment for an author who thought of himself primarily as a scientist and who had some difficulty in finding a home for his manuscript.

Why, among the less memorable works it so closely resembles, has *A Sand County Almanac* become such a favorite? Leopold has been justly praised for applying ecological principles to land-use planning and for developing the notion of a land ethic. But earlier scientists had introduced the term *ecology* in popular writings well before the turn of the century, and earlier writers, such as Muir and Thoreau, had suggested that humans should view themselves as members of a larger community of life. A student of history might conclude that Leopold merely popularized these ideas, pointing out that his work became widely known only during the late 1960s, in the heyday of environmentalism. But to a student of literature, these facts do not fully explain the book's lasting appeal.

To understand its sense of profound, unfailing richness, I believe we ought to examine *A Sand County Almanac* as a work of art. As with any piece of fine writing, we readers experience it in stages: we come with a set of generic expectations; we respond to the text episode by episode; and we form a sense of the work as a whole, both at the end of our first reading and whenever we return for another look. Regarding

Leopold's text in this light, we notice at the very start that he has chosen a form that both fulfills and challenges our expectations of what a nature book should be. Then, reading beyond the foreword, we see him engage our interest with a dense but transparent style and a charismatic persona. Finally, when we return to the book for a second or third reading, we realize that he has been writing in parables that force us into personal acts of interpretation. The book stays with us because, in reading it, we experience something analogous to what Leopold experiences in nature itself.

The term *nature writing,* which we would conveniently apply to *A Sand County Almanac,* usually means descriptive or narrative literature that falls between scientific reportage and imaginative fiction. This genre can include the most diverse subject matter without losing its formal distinctiveness. The critic John Hildebidle has described nature writing as "informal, inclusive, intensely local, experiential, eccentric, nativist, and utilitarian, yet in the end concerned not only with fact but with fundamental spiritual and aesthetic truths."[1] Nature writers rely upon scientific information and their own observations but feel free to explore these in a broader, more imaginative way than professional science would allow. Traditionally, too, nature writers have taken their inspiration from particular beloved places. In general, we might say that whereas scientists use experience and observation to generate "facts" (that is, theories), nature writers use scientific facts to enrich and deepen their readers' experience.

Although the roots of nature writing extend to such antecedent forms as the sketch, the letter, the diary, and the travel book, we can fix its debut at 1789, when the English clergyman Gilbert White published *The Natural History of Selborne.* This book, composed of White's letters to two fellow naturalists, established many of the standard features of the genre: reliance on scrupulous observation, reverence for fact, an affectionate regard for other creatures, an informal yet vivid prose style, episodic structure, diversity of material, and a charming local interest. White's book has gone through countless editions and has been repeatedly invoked as a formal and aesthetic standard by subsequent writers, including Emerson, Hawthorne, and John Burroughs.

During the century and a half between White's and Leopold's work, nature writing evolved considerably. The rise of romanticism led to an increased emphasis on the personality of the naturalist, particularly his emotional responses to the land and its creatures. Thus Darwin's *Voyage of the Beagle* (like White's book, a perennial favorite) not only gives meticulous descriptions of animals and plants but also records its author's outrage at slavery, his awe at the vast deserts of Patagonia, his exhilaration upon reaching the crest of the Andes, and his increasing wonder at the immensity of geologic time. Like White, Darwin rearranged and shaped the raw material of his diaries for publication, but he did so in order to present his voyage as an idealized circumnavigation where his geographic progress mirrors his developing understanding of nature.[2] Likewise, John Muir presents his ecstatic experiences in the High Sierra as examples of how people should relate to the land, and he depicts them as steps in the process whereby he himself was converted to "natural religion."

Romanticism also gave rise to an American tradition of social criticism based on the idea of nature's moral purity.[3] Early novelists like James Fenimore Cooper had imagined an American hero maturing under the virtuous influence of the wilderness, and painters like Thomas Cole and Asher B. Durand had depicted sublime landscapes enduring the rise and decay of civilizations.[4] But Thoreau was the most vigorous practitioner of this sort of cultural criticism, and his example, like White's in England, set a standard for all subsequent nature writing in America. Indeed, as I will suggest later, Leopold seems to have more in common with Thoreau than with any other nature writer.

By the mid-twentieth century nature writing was an established genre with a considerable history. No practicing naturalist would write without a high degree of literary self-consciousness, and no reader could approach a book like Leopold's without expectations of what he would find. These facts troubled the publishers to whom Leopold sent his manuscript. In particular, Leopold's correspondence with the editors at Alfred A. Knopf shows that they wanted a book of local observations and personal adventures, with ethical and ecological concepts mentioned only in passing. In other words, they wanted a Gilbert White book. But, as the letters show, Leopold felt that this model would not meet his needs.

Of course, *A Sand County Almanac* does answer many of our initial expectations. Like White, Leopold relies heavily upon his own observations, conveys a strong sense of place, responds with charming affection to the creatures he studies, writes in a vivid, unaffected style, and organizes his book episodically. Like White's romantic successors, Leopold foregrounds his own personality, indulging in social criticism as well as romantic effusions of praise and stressing the moral and aesthetic dimensions of what he sees.

However, Leopold's book exceeds our generic expectations in two important ways. First, though the title suggests otherwise, its scope is not local but continental. Leopold ranges from Sonoran deserts to arctic tundras with stopovers in the basin and range, the midwestern prairies, and the mountains of Arizona and New Mexico. Though the midwest gets the most coverage, all Leopold's descriptions convey an equally strong sense of place. The title of Part II, "Sketches Here and There," and Leopold's claim in the foreword that they are "scattered over the continent," suggest that these pieces are meant to be representative as well as particular. Thus, he addresses continental truths by means of local vignettes. Without sacrificing the strong sense of place, his natural history ceases to be "parochial" in the manner of White or Thoreau and becomes universal.

Second, as we learn from Leopold's foreword, this book will depart from the ostensibly casual and informal modes of presentation used by most earlier nature writers. At the very outset, Leopold makes us conscious of the deliberate artifice with which he has structured his text. The seasonal arrangement of sketches in Part I not only conceals their thematic connections but also reminds us of how carefully some of his predecessors organized their accounts. One thinks, for example, of the idealized year that Thoreau used to structure *Walden* or the idealized circumnavigation Darwin chose for his *Voyage of the Beagle*. Yet Part I more closely resembles White's brand of natural history than do Parts II or III. Part II, in fact, is written in the mode of confessional autobiography and recounts, in Leopold's words, "some of the episodes in my life that taught me, gradually and sometimes painfully, that the company is out of step" (viii). And Part III departs from natural history altogether by dealing with ecological and ethical issues in the abstract; it presents, in the most polemical terms, a program for social transformation. Thus, from the standpoint of form, the book appears to combine three separate genres, and we can understand why Leopold's editors were suspicious: by playing so boldly with his readers' expectations, he risked confusing or alienating them.

The foreword, however, does give a clue to his deeper intentions. The sketches of Part I, written for the most part in the present tense, describe the life he and his family live now. It sounds like an attractive life because it has a purpose (rebuilding what is lost elsewhere) and because it provides spiritual refreshment, "our meat from God" (viii). The sketches bear out these claims, depicting a life of wisdom, excitement, and intimate delight that any reader of nature books might envy. Part II, with its autobiographical episodes (written, for the most part, in the past tense), depicts the transformation of the narrator from an ignorant, insensitive, restless, and ordinary person (someone much like the reader) into the vibrant, perceptive, wise, and serene observer of Part I. Like all conversion stories, Part II inspires emulation. Thus, by the time we reach the abstract and programmatic essays of Part III, we are more ready to entertain their ideas and apply them to our lives than if we had simply read them first. The book's structure seems designed to create a climate of belief that will make us receptive to Leopold's doctrine of land citizenship.

Thus, *A Sand County Almanac* fulfills our generic expectations in some respects while challenging them in others. While it contains much natural history writing in the tradition of White and his successors, its purpose is not merely to delight and instruct. Like White, Leopold loves the creatures he studies and tries to explain their behavior, but, unlike White, he seeks actively to change his readers. *A Sand County Almanac* is a subversive book. It questions the deepest values of our civilization and challenges us personally on every page. In so doing, it allies itself more closely with Thoreau's method of social criticism based on the standard of nature. Yet Leopold differs from Thoreau by maintaining his role as a professional naturalist. He never lets his moral and social concerns carry him out of sight of the land itself.

When we get beyond the foreword, two features of *A Sand County Almanac* impress us at once: the brevity of its style and the personality of its narrator. From the first sentences of "January," one feels a curious tension in Leopold's prose. On the one hand, it flows smoothly and effortlessly. Each word drops into place with that sense of inevitability that Dylan Thomas said he found in all good poetry. The narrator's manner is confident and relaxed; his tone, though earnest, is rather light and conversational. He prefers a limpid, everyday vocabulary, avoiding jargon, scientific names, and verbal pyrotechnics. His words, in short, do not call attention to themselves. Nevertheless, one senses that each word carries a great deal of meaning, as if chosen with the utmost care. The smoothness and transparency of Leopold's prose belies its density. Like hand-rubbed wood, its surface conceals its craft. Viewed as a whole and compared to other works in the genre, his book achieves an epigrammatic conciseness reminiscent of the sermons of Jesus, the *Tao te-Ching* of Lao-tse, or Chomei's *An Account of My Ten-Foot Square Hut*.

Leopold succeeds here by using what students of short fiction call "techniques of compression." At one level, all storytellers face the same problem: how to present setting, characters, themes, and action in a way that holds the reader's interest and advances the plot. Novelists, blessed with a capacious form, can develop their characters, provide background, paint a scene, or complicate their plots at leisure. But short-story writers must squeeze their narratives into the space of a few pages without stinting on plot, characterization, or theme. Two principal strategies assist them here, and Leopold uses both. The first might be called "concentration": focusing the storyline on the climax and eliminating whatever does not advance the plot. The second technique might be called "engagement," a sort of narrative shorthand whereby the text invites the reader to supply vital information at which it can only hint.

Though Leopold's sketches are presumably factual, they are still good stories and thus have plots, characters, and themes. Like most nature writers, Leopold focuses on his interactions with other creatures, but unlike many of them he presents no fact that does not speak to the point. Whereas for someone like Darwin or White a natural fact may be intrinsically interesting, for Leopold such interest is not enough. The fact must serve the theme and thrust of the essay. Hence, Leopold's sketches, while conveying a good deal of knowledge, seem more coherent and memorable than those of most other naturalists. In his hands, natural facts take on a symbolic richness of implication, and in this respect he resembles both recent writers like John McPhee, a master of "symbolic journalism," and earlier writers like Thoreau, for whom natural facts always suggested higher truths. However, Leopold differs from Thoreau in that he never allows the moral theme to weaken his reverence for fact. Nor does he impose the theme upon the facts, but rather draws it out of them. In brief, he reads human life in the context of nature and not the other way around. Hence he avoids the sentimentalism of the popular writers whom John Burroughs scathingly termed "nature fakers."

Leopold's second technique of compression—what I have called "engagement"—invites the reader to contribute information that the text does not provide, thereby reducing the amount of explanation while increasing the density of implication. Leopold achieves this through repeated use of simple rhetorical figures, notably synecdoche, allusion, irony, understatement, and rhetorical questions. It is these "turns of phrase," rather than a self-consciously poetic vocabulary, that give his prose its memorable succinctness.

Synecdoche—the gesture of letting a part stand for a whole or an individual for a class—operates on the content as well as the form of *A Sand County Almanac*. In "January," for instance, the narrator confesses that his experience arises as a "distraction," suggesting that there is nothing special about it: it was not deliberately sought out, it could have happened to anyone, and it might well have included events other than the ones actually observed. In short, the experience is presented as both specific and representative. The same holds true for the episodes within the story itself. Leopold follows a skunk track, but he might just as well have followed the prints of a deer or a mink. On the way, he is distracted by signs of a field mouse, a hawk, rabbits, and an owl, but other creatures might just as well have appeared. These encounters illustrate general principles of ecology to the naturalist-narrator, whose acts of interpretation (given explicitly in the text) exemplify the sort of attitude we ought to take toward the land. Likewise, *Draba* stands for any "small creature that does a small job quickly and well" (26), the green lagoons of the Colorado represent any shining wilderness of one's youth, and Escudilla illustrates the sort of thing that might happen to any virgin ecosystem invaded by ignorant people. Each particular story invites us to imagine the same thing occurring in other regions. Synecdoche thus enables Leopold to limit the size of his book and restrict the contents of its episodes while at the same time addressing the most profound and universal questions.

Allusion also works to enrich the text, but does so by engaging the reader's knowledge rather than his imagination. Often a few key words will invoke a train of association that significantly affects our interpretation. In "January," for instance, Leopold notes that the skunk track runs "straight across country, as if its maker had hitched his wagon to a star and dropped the reins" (3). This image from Emerson's late essay "Civilization" invokes the Transcendentalist method of viewing nature spiritually.[5] Leopold will pick up that suggestion in "Good Oak," where he mentions the "spiritual dangers in not owning a farm" (6). But the image also tells us how to interpret the rest of this essay. To hitch your wagon to a star, as Emerson implies, is to let the heavens direct your economic life, and this, in a sense, is exactly what the creatures in a land community do. But the skunk also seems to be following some sort of celestial summons, as his odd emergence and abnormally straight track suggest. What does it mean, then, for the naturalist-narrator to follow this skunk? If he meets it, he may also learn something about the higher laws or divine powers that direct it. This winter walk is thus not merely a saunter punctuated with distractions, but a kind of vision quest. It is a quest, however, whose rewards come only from the distractions, since, when Leopold reaches the pile of logs into which the track disappears, he finds neither skunk nor star, but only the tinkle of dripping water that got him out of bed in the first place. The vision is withheld, and the naturalist turns home "still wondering." Yet, though his original questions remain unanswered, he has still found "meat from God" in the form of lessons in ecological interdependency and the perils of self-centered thinking.

Leopold uses another allusion to help us interpret these crucial "distractions." At the end of his encounter with the field mouse, he remarks, "To the mouse, snow means freedom from want and fear." Here he echoes Franklin D. Roosevelt's "Four Freedoms," connecting this apparently trivial incident to the entire context of contemporary politics: the depression, the New Deal, and World War II. We therefore realize that this is a tale not just of mice but also of men. Leopold personifies the mouse as a "sober citizen" and describes his "well-ordered world" in terms appropriate to a human metropolis, thus dignifying the mouse considerably while gently belittling human enterprise. In the end, however, the mouse's fatal self-centeredness stands as a warning to humans, who also place themselves at the center of things. Leopold's text invites us to see each world in terms of the other and thus gain a larger understanding. His allusion to the Four Freedoms challenges us to examine our own priorities and to entertain an ecological interpretation of history. Thus, he accomplishes here in only a few words what he will later take paragraphs to explain in his polemical essay, "The Land Ethic."

A third technique of compression uses rhetorical figures that invite the reader to say what's left unsaid. Irony, understatement, and rhetorical questions occur throughout the text. For instance, at the end of his ecstatic recollections of the Colorado delta, Leopold writes, "All this was far away and long ago. I am told the green lagoons now raise cantaloupes. If so, they should not lack flavor" (148). It is ironic, of course, that human beings would destroy an earthly paradise to grow breakfast foods, and Leopold's comment must be one of the bitterest in all literature. Similarly, in "Illinois Bus Ride," when Leopold passes a chanelled streambed, he remarks dryly, "The water must be confused by so much advice" (119). Further down the road he comes to a spanking-new farmstead where "even the pigs look solvent" (119). However, the fields are plowed right up to the fence, and the creek shows signs of flooding and gullying. "Just who is solvent?" he asks. "For how long?" (119). The alert reader quickly imagines the answers: not the pigs (who will soon be slaughtered) and not the farmer (whose soil will soon be eroded), but only the land itself, which is already being dissolved and carried away.

These techniques of compression both abbreviate and deepen Leopold's text, resulting in a peculiarly attractive style. Because the narrator does not browbeat us with verbiage, we feel respected, as if he valued our time, and so we are more inclined to listen. Here, we feel, is a writer who has taken pains to find exactly the rights words to express the distilled wisdom of his life. Such "verbal behavior" gives us important insights into his character and helps create the vivid narrative persona which so compels our attention on a first reading.

In any narrative, the reader must come to terms with the personality, values, and apparent intentions of the "implied author." This intimate relationship creates special problems for writers of confessional autobiography, who seek to justify their mature beliefs. They must affirm the views to which they have been converted, yet they must also convince their readers that the conversion was genuine, and that means helping their readers identify with their younger selves. Similarly, nature writers, who already love, understand, and appreciate other creatures, must win the sympathy and trust of their less sensitive and less well-informed readers. In both cases, the narrators seek to win their readers' trust and inspire emulation while at the same time challenging their normal behaviors and fundamental beliefs.

Leopold's naturalist-narrator gains our confidence in several ways. To begin with, he salts the text with references to his university, his graduate students, the jargon and nomenclature of his discipline, and the systematic field studies he has undertaken. Unlike White, Thoreau, and Muir, he bands chickadees and subjects migrating geese to statistical analysis: in short, he's a professional. But, like his predecessors, he is also well versed in literature, history, and philosophy, as we learn from his numerous allusions. He also relies heavily upon his own observations.

Indeed, he is as scrupulous and sensitive as White, with the same contagious curiosity, the same interest in field observation, and the same knack for making connections. Finally, as an accomplished hunter and outdoorsman, he seems to feel an emotional bond with his subjects different from what we find in Muir or Thoreau. The fact that he has killed woodcock, for instance, helps make the praise of "Sky Dance" not only more poignant but more profound. The essay is tinged with a darkness we have not encountered before, and this creates an impression that its wisdom is somehow more genuine.

But Leopold's narrator appears more than professional to us: we also find him an entertaining, engaging, and charismatic personality. We note at once his affectionate and sentimental regard for other creatures, an attitude quite reminiscent of White, Darwin, and Muir. Leopold loves Canada geese, sandhill cranes, and wild game of all kinds, yet he seems charmingly fond of small creatures that no one appreciates, like *Draba,* the chickadee, and the field mouse. His interest in undervalued and marginal things extends to landscape; he prefers the nondescript scenery of a sand county farm to the romantic sublimities of Muir's High Sierra or the edenic woods and pastures of White and Thoreau. Indeed, landscape as such hardly seems to interest him. What goes on *in* the land is what fascinates, and toward this he reveals an endearing capacity for the deepest feelings. His essays on wild geese, for instance, convey as much yearning, exultation, and praise as the most strenuous poem of Shelley. Leopold's sense of beauty is complex but it is romantic and involving rather than abstract and contemplative. Thus, it contributes to our sense of him as a warm and engaging person: we admire and are drawn to thinking people who can also be deeply moved.

Two other attractive features of Leopold's persona deserve special mention. He appears to us as a man whose convictions run as deep as his feelings. His long intimacy with the land seems to have bred both loyalty and a sense of moral responsibility. He considers it his duty to defend the land by bearing witness to the truth of which he has been convinced. The certainty with which he states his views, and the courage he shows in casting them before an uncaring public, create an aura of charismatic self-confidence. At the same time, he avoids melodramatic posturing by poking fun at himself from time to time. Thus, in "Good Oak," when the family goes back to sleep after a lightning strike, he remarks, "Man brings all things to the test of himself, and this is notably true of lightning" (8). And, in "Red Lanterns," he exchanges roles with his dog, whom he imagines treating him, the professional naturalist, as just another "dull pupil" (64).

Finally, Leopold's persona challenges or confronts us, even as it draws us out and wins our trust. This naturalist-narrator presents himself as an outsider, devoted to an unpopular cause yet convinced that "the company is out of step" (viii). Time and again he calls our attention to the glaring disparity between our view of the land and how the land actually behaves. At times he reacts with bitter sarcasm, as when, in "Good Oak," he quotes a Wisconsin governor's declaration that "state forestry is not a good business proposition" (10). Leopold does not call the governor a fool; he merely notes that "it did not occur to him that while the courts were writing one definition of goodness in the law books, fires were writing quite another on the face of the land. Perhaps, to be a governor, one must be free from doubt on such matters" (11). We can hardly help wincing for this poor gentleman. Leopold's dry wit reminds us of the opening chapter of *Walden,* where Thoreau adopts the same crusty manner (rather like that of a yankee farmer) to rouse his neighbors from their lives of quiet desperation. In both cases we, as readers, shrink from the object of scorn. Leopold's barbs entertain, but they also challenge us to examine our values. We would hardly like to have such a critic turn his wit upon *us.*

But this clever, accomplished, and formidable narrator is not always a happy one, and here again he challenges us. A deep current of melancholy runs through this book, even in the midst of its most rapturous celebrations. It soon appears that the narrator's conversion experiences have not really made his life any easier, but instead have made it more complicated and painful. To love a place is to suffer doubly when it perishes: witness Leopold's reaction to the loss of the green lagoons. To be able to read the book of nature means being able to see the ugliness and degradation as well as the beauty. Only Leopold notices the yellow cheat grass taking over the slopes of California foothills; only he knows why the cheat has replaced the more desirable native grasses. The price of his ecological wisdom is loneliness, isolation, and an aching sense of loss. If we wish to emulate him, we have to accept those terms.

To sum up, we may conclude that the brevity of Leopold's style reflects the values and personality of his narrator. Both serve the purpose announced in the foreword: to change the way readers conceive of and respond to the land. We might say that Leopold presents himself as a prophet, someone with special knowledge, a history of transformative experiences, and a "strange power of speech." Like the Old Testament prophets, Leopold finds truth in the wilderness and comes back to warn a society with little sense of its own spiritual danger. Like the New Testament prophets, he finds no honor in his own country, which certainly does not wish to be changed. Therefore, he resorts to the only weapons prophets have ever been able to wield: the strength of truth and the transforming powers of language. He takes his place with Thoreau as an American Jeremiah, judging his culture against the standard of wild nature.

Before we have gone very far in *A Sand County Almanac,* Leopold's sketches and arguments begin to make a cumulative impression. We finish the work convinced of its unity and purpose, and, when we return to it, we bring a clear sense of its moral vision. As with a great work of fiction, our knowledge of the ending only increases our delight in the storyteller's art. Here, understanding the doctrine of the land ethic only makes Leopold's vignettes more resonant.

A Sand County Almanac has more thematic unity than many classics of natural history. White's *Natural History of Selbourne* may demonstrate a method, but it hardly advances a philosophical or moral position; what unity it achieves comes from the personality of the narrator and the narrow range of his excursions. Similarly, Darwin's *Voyage of the Beagle* gathers the most diverse researches and concerns under the umbrella of an idealized chronology and the viewpoint of a single narrator; it has, in this sense, little more unity than a picaresque novel. *A Sand County Almanac,* in contrast, presents its sketches as moments of insight that point toward a core of truth the way iron filings respond to a hidden magnet. Only upon rereading the book do we fully appreciate how much its stories and underlying themes reinforce each other. Eventually, we realize that Leopold's "sketches" are really parables and that his parabolic style accounts more than anything else for the book's perennial freshness.

A parable conveys novel ideas by means of familiar facts and situations, as in the well-known parables of Jesus. The story of the mustard seed and the tares (Matthew 13) uses agricultural imagery to teach celestial truths. Jesus explains that the kingdom of heaven is like "the least of all seeds," yet when it is grown, "it becometh a tree, so that the birds of the air come and lodge in the branches thereof." Likewise, he compares God to a farmer whose enemy sows tares among his wheat, thinking to ruin the harvest; but the farmer waits till the crop is full grown and then has his servants bind the tares and cast them into the fire. Both scenarios must have been quite familiar to Jesus' audience, yet to interpret the parables one needs a rudimentary sense of Christian doctrine, as we learn a few verses later when Jesus' dull disciples ask for an explanation. To interpret, one must first believe, or at least entertain the possibility of belief. Then, the act of interpretation can give the doctrine new meaning through application to our life. But this application familiarizes the doctrine and so increases our incentive to believe. Thus, the parable initiates a spiraling process of interpretation and belief (sometimes referred to as the "hermeneutic circle"). Presenting a truth in this way, rather than by argument, has the advantage of winning from the reader not only understanding but commitment.

The "doctrine" behind *A Sand County Almanac* might be said to rest on three central ideas, which Leopold states in his foreword. These are that land is a community, that land can yield a cultural harvest, and that land should be loved and respected. These ideas are conveyed by three archetypal images occurring throughout the text: the analogy between natural and human systems (such as economics or politics), the metaphor of nature as a language or system of signs (the "book of nature"), and the figure of the naturalist-narrator as an example of how humans ought to relate to land. All three emerge clearly in the first two essays, which thus provide a key for reading the rest of the book.

In "January" the narrator's main discovery is that both natural and human communities are governed by similar economic concerns. The narrator shows enough love and respect for the skunk to follow his track, leave him alone in the driftwood pile, and write a charming essay about him later. In "Good Oak" the narrator's main concern is reading the book of nature, which he does by examining the traces of recent activity (rabbits, present seedlings, squirrels planting acorns) and the history of the land as recorded in the chronicle of the oak tree itself. Reading the book of nature is thus both investigative ("reading sign") and interpretive (reconstructing the past). The naturalist expresses his love and respect not only by referring to the oak in honorific terms ("dead veteran," "no respecter of persons," "what is a ton of ice, more or less, to a good oak?") but also by ritualistically returning its ashes to the soil so that they may rise to new life in another form.

To illustrate how the parabolic style works, let me conclude with a look at "Thinking Like a Mountain." Leopold once considered this essay for his title piece, and certainly it is one of his most memorable.

The title, at first glance, appears cryptic and paradoxical, much like a Zen koan: how could a huge, inanimate object "think"? The answer does not emerge till the end, and, like the solution to a koan, it requires a fundamental change in the reader's world view. The essay points the way for this change by presenting a story of conversion and inviting us to identify with the narrator. In this sense, it reflects the ultimate concern of the book as a whole.

"Thinking Like a Mountain" begins with a vivid evocation of a wolf's howl. Imagined, not recounted, and put in the present tense, this opening gives the reader an immediate vicarious experience. The next two paragraphs state the theme: how to "decipher the hidden meaning" of that call, or, by extension, how to read the signs in the book of nature. Paragraph two presents a series of misreadings, much like those of the creatures in "January" who misunderstand why snow falls or grass grows. Only the mountain, it seems, can interpret the call "objectively," avoiding the fatal errors committed by self-centered beings with a limited sense of history. By the end of the third paragraph, then, we already sense the moral of the conversion story we are about to hear: that reading the book of nature will yield a "deeper meaning" and put us in touch, through a shared point of view, with the mountain itself.

Paragraphs four through six shift to the past tense, presenting the experience that converted the narrator to his present practice of thinking like a mountain. It is a brief, vivid, and intensely dramatic tale with a violent climax and a mournful, repentant ending. As he watches the green fire die in the old wolf's eyes, the narrator feels his truimph turn to a horrified realization of sin. But his sin is explicitly described as an act of misinterpretation: "I realized then, and have known ever since, that there was something new to me in those eyes" (130). Up to this point, the narrator's understanding has been as limited and subjective as those of the creatures he lists in the second paragraph. But he regrets it now: "I was young then, and full of trigger-itch" (130). His retrospective view comes with the bitterness of tragic insight. He now recognizes the consequences of misreading nature's signs: the murder of a fellow-creature, and his own alienation from a greater sentient being (the mountain itself).

The second section of the essay, divided from the first by asterisks, describes later experiences that confirm the truth glimpsed at the moment of conversion. Here the narrator bears witness to the worth of his new belief. Significantly, the encounter with the dying wolf has empowered him to read the book of nature correctly. He notices the slopes of other mountains "wrinkle with a maze of new deer trails" (130) once the wolves have been killed. He can explain the bleached deer bones and denuded shrubs of the dying range. But he can also read the history of the land and predict its future, because he can understand the rules by which the signs he reads are produced. That is, he can think ecologically; he can think like a mountain. In the second paragraph we learn that the term "mountain" means not just a pile of rock but the entire community of creatures who live there. "Mountain" is thus synonymous with "range," and "thinking like a mountain" means considering the needs of the whole community rather than those of one member alone. The third paragraph extends this insight to cows and ranchers, warning that the latter's own chronic misperception will one day prove fatal: "Hence we have dustbowls, and rivers washing the future into the sea" (132). This final declaration shows that the narrator's conversion experience and subsequent correct interpretations have turned him into a prophet, thus fulfilling the role he assumed in the foreword.

The third section of the essay presents what might be called the moral. Here the narrator steps back from his experiences and predictions, discarding the intense voices of penitent and prophet to assume a serene and reflective tone. He ponders the deeper connection of ecological principles and events to our own moral and social life. His scornful allusion to Neville Chamberlain's slogan "peace in our time" suggests a link between incorrect interpretation and moral laxity: we deliberately misread because we want to feel safe. But too much safety, as the starved deer know, may lead to death in the end. The essay concludes by invoking the wisdom of Thoreau, our foremost prophet-naturalist, but in such a way that we must construe his words from a new ecological viewpoint. The final sentence returns to the image of the wolf call with which we began and once more raises the question of "hidden meaning." Though we cannot be sure of its message—as the narrator's "perhaps" reminds us—we are a lot wiser (and sadder) than we were at the outset. Now the title no longer seems paradoxical: now we can imagine what it *could* mean to "think like a mountain." Thus the essay comes full circle, beginning and ending with the problem of interpretation. Its moral, I would suggest, is that everything we thought we understood—the wolf, game management, the raising of livestock, our laws, our politics, even the words of our prophets—must now be reviewed from an ecological perspective. Thus, the structure of the essay reflects the parabolic movement of interpretation itself, and, in so acting to transform our consciousness, it serves as a paradigm for the effect of the book as a whole.

A Sand County Almanac has become a classic because of the ways it arouses and holds our interest: by playing upon our expectations of what a nature book should be, by creating a dense but transparent style and an attractive but challenging narrative persona, and by conveying its wisdom in parables that change our angle of vision. To call it a classic, however, is not to say it is perfect, merely that it endures. It rewards rereading with increased delight and deeper, more personal instruction. Like Keats's Grecian urn, it works to tease us out of thought and into imagination, not by virtue of its contents, but by the manner in which they are conveyed. That Leopold had a rare literary gift cannot be doubted, and one wonders what more he might have written if he had lived longer. But what can one add to the distilled wisdom of a lifetime? The gospels, too, are brief, challenging, and wonderfully durable. We could ask the same question of them: not, what more do we want, but what else do we need?

NOTES

1. John Hildebidle, *Thoreau, A Naturalist's Liberty* (Cambridge: Harvard University Press, 1983), 61.

2. See John Tallmadge, "From Chronicle to Quest: The Shaping of Darwin's *Voyage of the Beagle*," *Victorian Studies* 23 (Spring 1980): 325–46.

3. See Perry Miller, "The Romantic Dilemma in American Nationalism and the Concept of Nature," in *Nature's Nation* (Cambridge: Belknap Press of Harvard University Press, 1967), 197–207.

4. See Miller, "Romantic Dilemma," 197–98; also Barbara Novak, *Nature and Culture* (New York: Oxford University Press, 1980).

5. Ralph Waldo Emerson, "Civilization," in *Society and Solitude* (Boston: Houghton Mifflin, 1904), 30.

Barry Lopez
1945–

> If it is true that modern people desire a new relationship with the natural world, one that is not condescending, manipulative, and purely utilitarian and if the foundation upon which the relationship is to be built is as I suggest—a natural history growing out of science and the insights of native peoples—then a staggering task lies before us.
>
> —Barry Lopez, *Crossing Open Ground*

Barry Lopez was born in Port Chester, New York, in 1945 but spent his early years in rural southern California. His family moved back east to New York City when he was ten, and he spent the remainder of his early years there. He attended prep school in the city and received his B.A. from Notre Dame in 1966. After college, he returned to the West, married (in 1967), and began to prepare himself for a career as a teacher and scholar. He received an M.A.T. from the University of Oregon in 1968 and continued to do graduate work there until 1969, when he decided he would rather be a professional writer. He has lived in western Oregon since 1968 and has been a full-time writer since 1970.

Lopez's first book, *Desert Notes: Reflections in the Eye of a Raven*, was published in 1976 and consists of eleven short fictional narratives. Since that time, he has published a collection of American Indian trickster stories, two more collections of short fictional narratives, an Indian animal fable, and four books of nonfiction: *Of Wolves and Men*, *Arctic Dreams*, *Crossing Open Ground*, and *The Rediscovery of North America*. These four works of natural history (as he likes to label them) are largely responsible for Lopez's prominent place among contemporary American writers, though critics have recently begun to pay more attention to his unusual fictional works.

Lopez has been a great success as a writer almost from the beginning of his career and has received many awards and honors for his work. *Of Wolves and Men* (1978) was immediately recognized as an outstanding work of natural history. It is carefully and thoroughly researched, exceptionally well-written, and—among other things—a searching and disturbing examination of the relationship between humans and wild nature in the West, as exemplified in our relentless extermination and demonization of wolves. *Arctic Dreams: Desire and Imagination in a Northern Landscape* (1986) is the most comprehensive and complete example, to date, of Lopez's range and power as a writer and of the compelling moral vision that underlies (drives) his work.

Most of what Lopez has written (fiction and nonfiction) is concerned with the different kinds of relationships humans now have, have had, and need to have with nature, especially in the New World, and particularly in the part of it that Lopez knows best, North America. Somewhere along the way, we—the human inhabitants of the West—made a wrong turn and lost our way in our relationship with nature, so more and more we are less and less at home in the natural world and ever more exploitive, destructive, and economically motivated in our relationship to and treatment of nature and the wilderness that was here when we arrived. Much of what Lopez has written is strongly affirmative and seeks ways in which we may refind the way that was lost, or, more accurately, ways in which we may find a new way that will teach us to be at home in the natural world, by cultivating tolerance and respect for all other life forms and developing a set of human values that will mediate between the extremes of civilization and nature so that humans may find a dignity that includes all living things.

Barry Lopez and the Search for a Dignified and Honorable Relationship with Nature

William H. Rueckert

1

Since he began publishing books in 1976, Barry Lopez has written in three different modes. First is the metaphorical, fictional, narrative mode of *Desert Notes* (1976), *Giving Birth to Thunder* (1977), *River Notes* (1979), and *Winter Count* (1981)—all short books consisting of brief, usually unconnected pieces. Only *River Notes* tells anything resembling a consecutive narrative. The Coyote stories in *Giving Birth to Thunder* are all connected in one way or another. Second is the research, historical, personal experience mode of *Of Wolves and Men* (1978) and the monumental *Arctic Dreams* (1986). And third is the journalistic mode of his many essays, which often also make use of or are based in firsthand personal experience, as represented in his recent, carefully selected collection of essays *Crossing Open Ground* (1988).[1]

Lopez calls himself a journalist, which is somewhat misleading, because he is also a masterful storyteller and a careful, even meticulous researcher. Everything he has published is deeply moral (as opposed to the so-called objectivity of the journalist) and profoundly based in an ecological vision, itself based in a somewhat spiritual, pre-Western view of wilderness and landscape and of man's relationship to nature. A

The original version of this essay was written in 1989 and published, under the same title, in a special issue of the *North Dakota Quarterly* 59 (Spring 1991): 279–304, devoted to nature writing. This is a shortened, revised, and partially updated version of that essay.

powerful nostalgia for the primitive runs through most of his work and takes the form of a desire to return to, or, more realistically, to retrieve, the kind of intimate, reciprocal relationship hunter-gatherer cultures had with nature, a relationship based on an extensive firsthand practical knowledge of nature and a reverential, nonadversarial attitude toward it. This relatonship is best exemplified for Lopez in the Eskimos and the North American Indians before their way of life was destroyed or corrupted by Western culture. This position is carefully laid out and discussed in many of the essays in *Crossing Open Ground*, in numerous passages throughout *Arctic Dreams,* and in the purpose behind the writing of *Of Wolves and Men.* It is also embodied in many of the stories in *Desert Notes* and *Winter Count,* and in the whole of *River Notes*—the best of his books in the metaphorical, fictional, narrative mode. Of all his books, *River Notes* best exemplifies the importance Lopez attaches to the role of imagination and metaphor in writing and to the educational role of the writer in human affairs.

Lopez is never frivolous. From the very beginning, he has been a writer with a mission. This is probably most obvious in such books as *Of Wolves and Men, Arctic Dreams,* and *Crossing Open Ground,* where the educational mission—the didactic intent—is often stated very clearly. Anyone who pays attention to ecological matters knows that we humans—especially we Western humans—with our advanced science and technology and with our obsession with the economic motive, are destroying the earth, ravaging nature and the other life forms that cohabit it with us, and breaking the ozone layer. This knowledge is hardly new, and the need for a new land or nature ethic, a new attitude toward and set of values for the nonhuman, existed long before Leopold formulated and named it in his now justly famous essay.

All Lopez's major works originate from an awareness of this need to reestablish a more harmonious, less destructive, reciprocating, respectful relationship with nature—with the prehuman and nonhuman ground of all life; with the preverbal, nonverbal, and pretechnological. It is a relationship with the Other—not the alien, and not the divine or transcendental, but the wild, the wilderness, the most varied gene pool of all life. In our obsession with the rational, with order, with simplifying and bringing under human control, we humans are systematically conquering, domesticating, destroying, reducing, and simplifying this other in the process of putting it to exclusive human use.

As Lopez knows, and as he shows in *Of Wolves and Men, Arctic*

Dreams, and *Crossing Open Ground,* our attitude toward and relationship with nature is usually grounded, these days, in economic, religious, political, nationalistic, and purely self-interested motives. The whole history of human civilization almost everywhere has been the history of the destruction of the wilderness or of the transformation of the natural environment from the prehuman and nonhuman to the humanized, as epitomized in all of the great cities of the world and, in lesser ways, in all human settlements of any size. One reason Lopez admires the Eskimos and the North American Indians is that they lived in the natural environment and learned how to use it to their own ends without destroying it. Many were hunters and gatherers and were preagricultural and pretechnological; few had a written language. The were often nomadic; most did not build cities. Even under the harshest conditions, as with the Eskimos, they lived in harmony with nature, unlike the oil exploiters and workers Lopez describes with such savage irony in *Arctic Dreams* or the wolf haters and killers of all ages described in *Of Wolves and Men.*

With the exception of "The Bull Rider" (1978) and "Grown Men" (1978), the two earliest and somewhat atypical essays in the book, the carefully selected essays in *Crossing Open Ground* are works of the eighties in which Lopez explores, examines, and evaluates his and other humans' relationships with and to nature. His recurrent concern is our present need to get in touch with (or back in touch with) nature (the nonhuman ground of being) so that we can come to know, learn to appreciate, and draw on its wisdom. Most of the essays in *Crossing Open Ground* use narrative as the main organizing device, as opposed to a discursive or argumentative or thesis-centered form of organization. Most of the essays tell a story, sometimes organized around something that happened to Lopez: a trip into the desert to see the stone horse, a trip to Tule to see the snow geese, a trip down the Colorado River, or a trip down the Yukon-Charley River. A few of the essays—such as "The Passing Wisdom of Birds"—are loosely organized around a sequence of events— here Cortes's conquest of Tenochtitlán and the burning of the aviaries— but the narrative in these essays is subordinated to the thesis of the essay and the overall thesis of the book. There are no fictions—as far as one can tell—in this book, though there are many narratives and lots of metaphors—or, perhaps more accurately, lots of symbols and symbolic actions. Cortes's burning of the aviaries is a symbolic act; the stone horse is a symbol; and going to see the thousands of snow geese and traveling down the Colorado River are symbolic actions. Lopez is seldom, if ever, interested in what we might here call pure "nature writing"—that is, describing nature, or one of his experiences in nature (such as going down the Yukon-Charley River) for its own sake, as a way of celebrating nature—the way Thoreau does in his journals. He is a superb writer, so there is much wonderful nature writing in most of his books. An excellent example from *Crossing Open Ground* is "Trying the Land."

Lopez's real subject everywhere is not nature writing for its own sake or nature for its own sake (as we get it in some of the best nature writers) but man's attitude toward and relationship to nature and the urgent need to break both free from the exploitive and destructive Western view that has prevailed for hundreds of years. Lopez argues in his recent books that this view has had calamitous effects in the New World. It has led to the destruction of much of the original wilderness in North, Central, and South America, and if it is left unchecked and unaltered, it may well soon complete its destructive work, as advanced technology makes it possible to exploit natural resources at an ever faster rate, even in such remote and inhospitable environments as the Arctic. What is possible in the face of this seemingly irreversible exploitation and destruction of wilderness and the wild creatures that inhabit it in the name of "progress" and economic growth? Some of the essays in *Crossing Open Ground* and the more recent *The Rediscovery of North America* explore both the causes and effects of this Western view of nature and propose alternatives. If Lopez sometimes seems a bit naive in his hopefulness and proposals, it is perhaps a function of his age (he was born in 1945) and his determination to avoid cynicism, despair, and the futility of historical irony.

Lopez is concerned with what we have lost—a kind of relationship to wilderness (including wild animals) that he feels is essential to the good life—and with how we need to change if we are to regain the attitudes necessary to this kind of relationship to wilderness—that is, to the not yet humanized, to the Other, to the prehuman, to the ultimate ground of all being, to the open ground so suggestively referred to in the title *Crossing Open Ground*. We have to experience the wilderness, acknowledge the reality of it, and allow it to enter us and have its rightful place in our total view of reality. This has been Lopez's main concern since at least *River Notes: The Dance of Herons* (1979), and it continues to be his primary concern in his recent book, *Crossing Open Ground* (1988). I want to examine *River Notes* in some detail, because using the freedom that fiction, imagination, and metaphor allow him,

Lopez is able to examine different relationships between humans and nature (the not human) in very subtle and suggestive ways, and to make clear to anyone who takes the trouble to read this wonderful book with the same care that Lopez took in thinking it out and writing it, what he thinks a healthy, generative attitude toward and relationship to nature must entail.

2

River Notes is a short (eighty-one pages), dense, sometimes puzzling book consisting of a brief introduction and eleven unnumbered notes, the average length of which is about four pages. The introduction and the majority of the notes are in the first person, though not all are directly about this "I." The introduction begins on an ocean beach where the river ends, with a man (the "I") waiting for something. He has been waiting for a long time. You do not know what for, but he does. His actions are rather strange and forewarn us that things may not always be ordinary and completely realistic in what is to follow. *River Notes* is not just another naturalist's observations of a river.

The titled sections begin and end with episodes involving the river and herons. Why herons? Lopez obviously likes herons. We do not know why—in other words, how or why he has personalized them and made them symbolic—until we finish the book and come, at last, to the dance of the herons in the last note, "Drought." But he does not begin with the herons; he begins with the death of a river (when it loses itself by flowing into the sea), just as he ends with the beginning of a river, or, more exactly the rebeginning of a river after a long, terrible drought. He begins passively with the "I," waiting and watching; and he ends actively, when an act of the "I" toward nature (he saves a fish stranded and dying because of the drought) causes the dance of the herons, which brings the drought to an end. The ending is active in another sense as well; the herons teach the "I" to dance because of his selfless act, and he participates in it with them. At the very end, the rains begin, and the river comes back to life.

> The river has come back to fit between its banks. To stick your hands into the river is to feel the cords that bind the earth together in one piece. The sound of it at a distance is like wild horses in a

canyon, going sure-footed away from the smell of a cougar come to them faintly in the wind. (81)

The book begins:

> I am exhausted. I have been standing here for days watching the ocean curl against the beach, and have sunk very gradually over all these hours to the sand where I lie now, worn out with the waiting. At certain moments, early in the morning most often, before sunrise, I have known exactly what I was watching the water for—but at this hour there is no light, it is hard to see, and so that moment passes without examination. (ix)

He is in nature, waiting for some event in nature to occur. His thoughts are somewhat strange, as are his actions. At one point, right after the beginning, while he is lying in the sand, he says, "I reach out and begin to dig in the sand, feeling for substance, for stones in the earth to hold onto: I might suddenly lose my own weight, be blown away like a duck's breast feather in the slight breeze that now tunnels in my hair" (ix). He tells us that he has been "alert to the heartbeats of fish moving beyond the surf" (x) and that he has "spent nights with [his] hands flat on the sand, tracing the grains for hours like braille until [he] had the pattern precisely, could go anywhere—the coast of Africa—recreate the same strip of beach down to the very sound of the water on sea pebbles out of the sound of [his] gut that has been empty for years; to the welling of the wind by vibrating the muscles of [his] thighs" (x) He says that he "knew the sound—of fish dreaming" and that he "dreamed" he "was a salmon" (x).

A bit later he says, in this same vein—that is, speaking always about his relationship to nature:

> I hold in my heart an absolute sorrow for birds, a sorrow so deep that at first light of day when I shiver like reeds clattering in a fall wind I do not know whether it is from the cold or from this sorrow, whether I am even capable of feeling such kindness. I believe yes, I am. (xi)

Then he describes an episode when, one rainy winter dawn, standing on the beach, he felt birds begin to alight on him as if he were a tree: golden plovers, black turnstones, red phalaropes, murrelets, sanderlings, whimbrels, avocets, and finally eider ducks. As the weight of all these birds brings him to his knees, he says he "could feel such anguish as must lie unuttered in the hearts of far-ranging birds, the weight of visions draped over their delicate bones" (xi). He goes to sleep. When he awakens the next day—still on the beach—he washes himself in a freshwater pool formed by a river that "broke out of the shore trees, ran across the beach, and buried itself in the breakers" (xii). He makes a kind of soap out of talum roots and washes the ashes of last night's fire from his hands and the fear of darkness and death from his mind (xii).

The action of entering the river is repeated throughout the book, by the "I," other humans, and various nonhuman things and creatures. For the humans, anyway, the experience of entering the river is always charged with significance. I avoid the word *symbolic* here because there are very few systematically developed symbols or symbolic events in the book. If the river, for example, were interpreted as being symbolic of nature rather than the living organic reality itself, the significance of the whole book would be lost. More than anything else, this book tries to destroy the human tendency to reduce nature to or transform it into something that it is not, especially abstractions, symbols, or formulae. Nature is. Birds are. Herons are herons. Stone is stone. Grains of sand must be experienced, known, and understood in terms of their own being and individuality. The river is there to experience, know, share being with. The same is true of nature in the larger sense. If you try to reduce very much that you find in Lopez's book to symbols, you will only stub your mind.

At the very end of the introduction, the man who has just entered the river has eschatological thoughts about the ends and beginnings of rivers and about the last days that are sure to come for all of us (every individual living being, including the river) there where the river and sun run into the ocean. And there where the river ends, where it "buries" itself in the sea, he thinks about going upriver, to the headwaters, and beginning again. The introduction ends with a direct address to the reader ("you") telling us what we will find if we follow the river into the trees, up its whole length, to where the headwaters are and where the river begins its long journey toward death in the sea.

We never know exactly who this "I" is or, really, what he is waiting for there on the beach. But we do know that the "I" tells bits and pieces of "his" story and bits and pieces of the river's story. The

"I" is the eye of this whole story, the receiver and observer, the actor (the one who acts), and the recorder of these *River Notes*. The "I" is also the person who dances with the herons and is directly involved (as actor or observer) in a variety of other strange actions chronicled in these notes. There are eleven short sections following the introduction, all of which are located on and are directly about the river. In the next section, "The Search for the Heron," we move from the beach to the river. Here the "I" observes the herons and tries to learn about them. Most of what he discovers is told to him by the trees and by the herons. The knowledge he acquires is always double, even triple: about the herons, about nature, and about himself, reality, and human conduct. As he does throughout the book, he enters the river in an action that is always vaguely ritualistic and symbolic.

In the next section, "The Log Jam," are notes arranged chronologically from 1945 to 1973, all of which have to do with something or someone entering the river. In the fourth note (1957), a strange episode in the life of a lonely woman is described. In a moment of extreme loneliness and sorrow, while her husband sits reading, she takes a bowl full of old flowers, goes to the river, removes all of her clothes, and walks into the cold river throwing the flowers into the river until the bowl is empty. She thinks about but does not commit herself to the river. In the final note, we are told of a boy who lives beside a fallen tree and has the same kind of strange modes of perception into nature that the "I" does. He can put his forehead to a tree and hear the "thinking of the woodpeckers" inside. He can hear the flow of sap. Before the great tree fell, "he had heard the slow movement of air through the lengthening termite tunnels and had known" (18-19).

This book, we slowly come to realize, is about a need to have a deep extraverbal or nonverbal knowledge of nature that makes it possible to enter into a profound nondestructive relationship with all parts of it. In the previous section, "The Search for the Heron," the "I" decides that he may have this kind of knowledge of the heron and that because of it they will "dance together some day" (4). But he still has doubts: "Before then will I have to have been a trout, bear scars from your stabbing misses and so have some deeper knowledge? Then will we dance? I cannot believe it is so far between knowing what must be done and doing it" (5-6). When he learns about the heron's dreams from the cottonwoods, he decides he has made an "unholy trespass" (6). The next section of the book, "The Bend," is about this same drive to know and focuses on the questions that the "I" continuously raises about how one knows the absolute other, the nonverbal; about the relationship between knowledge, power, and destructiveness; and about how one can enter into a generative relationship with this absolute other—here nature, the bend in the river.

"The Bend" is a turn in the river the "I" has long wanted to "take the measure of," "to grasp for private reasons" (23). He wants to "wrestle meaning" from this spot (24). These terms of acquisition are strident and aggressive. The section begins with a denial of such terms and then goes on to tell us how the "I" learned to deny and relinquish this typical western attitude toward the relationship with nature. If we have read Merwin and Snyder, we think of their concerns about self and other, selflessness, transcending the obsession with self through a knowledge of and relationship to nature, and learning that it is enough just to know without having to use or exploit.

"The Bend" begins at the end of what has happened, retrospectively.

> In the evenings I walk down and stand in the trees, in light paused just so in the leaves, as if the change in the river here were not simply known to me but apprehended. It did not start out this way; I began with the worst sort of ignorance, the grossest inquiries. Now I ask very little. I observe the swift movement of water through the nation of fish at my feet. I wonder privately if there are for them, as there are for me, moments of faith. (23)

Then he tells us some of the things he has learned about the bend in the river, how he came to know some bends, and how, as he learns more, he feels "closer" to the bend. What he learns about is the ecology of the bend in the river. He explores the river with his hands; he has "eyes in the tips of his fingers." He knows the life along the banks of the bend and which rocks are gripped by slumbering water striders; he knows where beneath the water lie the slipcase homes of caddis fly larvae (23–24). As this kind of knowledge increases, he says that he feels he is coming closer to the bend—not in proximity but in ontology, the being of the bend itself. Then he makes a wonderful statement that begins to deliver the being of this book to us and allows us to get closer to it: "For myself, each day more of me slips away. Absorbed in seeing how the water comes through the bend, just so, I am myself, sliding off" (24). He wants to experience and know the bend in the river the way he does the herons, so that he can share being with it and dance with it.

His first attempt to wrestle meaning from the bend in the river was very different, and the account of it must be read carefully. He becomes ill and is eventually confined to his bed. Unable even to imagine himself well, he becomes obsessed with the bend in the river: "If I could understand this smoothly done change of direction I could imitate it, I reasoned, just as a man puts what he reads in a story to use, substituting one point for another as he needs" (24). So he sets out to acquire this understanding by "measuring" every part of the bend in the river. He becomes "obsessed with calculation" and calls on surveyors, geodesic scientists, and hydrologists. They do the measuring for him, over and over again, because he is convinced "that in this wealth of detail a fixed reason for the river's graceful turn would inevitably be revealed" (25). Lopez makes two points here: one has to do with the reduction of nature to scientific data, and the other has to do with the separation of the self from any direct sensory experience of nature for its own sake.

The scientists finally reduce the "bend in the river to a series of equations," which they take to the sick man, along with all of their notebooks, which he has piled in a corner of his room. He thinks that his problem has now been solved by these equations, which would provide him with an "incontestable analogy" on which to base his own change of direction and so get well again. But he is mistaken. In a wonderful series of transformations, the pile of notebooks first turns into water, and mergansers "suddenly fly off" from them. Moss grows on the notebooks. They begin to harden and resemble boulders in the river. "Years pass," and his depression goes on.

Suddenly, without warning, his depression begins to break one morning. He leaves his room, calling out for help. Bears hear him and carry him to the river, as he requests. The first thing he does is "feel, raccoonlike, with the tips of his fingers the soil of the bank just below the water's edge." He listens "for the sound of the water on the other bar." He observes the "hunt of the caddis fly" (25–26). "I am now," he says, "taking the measure of the bend in these experiences" (25). The last image in this section—a beautiful and characteristic one in this book—is how, to a hawk looking down on the bend, he (the "I") would appear for the moment to be "inseparably a part [of it], like salmon or a flower" (26). "I cannot say well enough," he tells us, "how this single perception has dismantled my loneliness" (26). And he tells us, after the fact, that his new experiences with the bend have caused him to lose "some sense of" himself and that he no longer requires as much.

This loss or diminishment of the individual ego, of the I-self that is going to discover the secret of the bend in the river by reducing it to scientific data and elegant equations and then putting it to human use, is a good and necessary loss in this book. Power over nature, usually based on knowledge of nature, is one of the most persistent (and often destructive) of all human and scientific drives. The domestication, exploitation, manipulation, and conquest of nature is probably one of the most characteristic of all human stories. There would be no human history without these motives. They provide all ecological literature with its basic subject matter and themes. The adjustment of human drives and needs to the laws and needs of nature is the most basic and pressing problem of our time. Barry Lopez has written about this, brilliantly, in *Of Wolves and Men* and *Crossing Open Ground*. In *River Notes*, his concerns are somewhat different, though related. This is an intensely personal, highly subjective book about one man's need to be in direct sensory contact with "wild" nature and the nonhuman, nonverbal world. But it goes beyond mere experience: the "I" of the book wants to know and understand this natural world without trespassing on it or causing it any harm. And he needs to be a part of it, to be accepted by it, so that he can dance with the river and the herons. We are reminded of Ike McCaslin's needs in the "Old People" and "The Bear," especially in the great episode in "The Bear" where Ike relinquishes his watch, gun, and compass and goes into the wilderness alone to meet Old Ben face-to-face on the bear's own ground. We are at a level of perception and knowledge here that is deeper than, before or beyond, reason. It is rendered in words because there is no other way to render it, though much of experience is nonverbal, even preverbal. To speak of primitivism here is not useful, because Barry Lopez and Faulkner are not primitives; nor, I think, would they have wished to be. It is useful, though, to speak of connectedness and community in the largest possible sense—of the biosphere or ecosphere or planet earth. With this concept, we come to the matrix of *River Notes* and begin to understand the significance of the subtitle—*The Dance of Herons*—and the fact that the "I" must perform a selfless act toward nature that earns him the right to join the dance of the herons, a communal event joining human and nonhuman, which brings the rains and renews the river and all the life in it and along its banks. Many of us wish that we could dance with the herons, and this book taps and releases those desires in profound and moving ways.

In the next section of the book, "The Falls," the "I" tells the story—the legend?—of a nameless friend who, though he does not actually dance with the herons, lives in perfect conjunction with both the human and natural worlds. He is able to live and be in both worlds. This Proteus-like figure is, at different times, a boy, a dog, a man, a fish, a salmon, the falls, the wind, a bird, and a bear. It is not always easy to tell who or what he is, the source of his power, or the reasons for some of his strange actions. He is an extraordinary figure who comes into and goes out of the narrator's life at infrequent intervals—three years, ten years, six years, ten more years—wandering and working all over the Midwest and Northwest. He is always alone, except when he is with the narrator. We hardly know what to make of his actions: changing places with his dog; becoming the wind or a bird; dreaming for four continuous days and nights; choppping a hole in the ice and bathing in the freezing river, then lacerating his arms in a strange bloody ritual and crying out like a bear; scraping his whole body with a knife and leaving the scrapings scattered about when he leaves the desert; brushing the body of the narrator with cuttings from ash and cottonwood trees; and finally leaving the human world, thus, recalling the picture on the book's cover.

> I saw him all at once standing at the lip of the falls. I began to shiver in the damp cold, the mist stinging my face, moonlight on the water when I heard that bear-sounding cry and he was shaking up there at the tops of the falls, silver like a salmon shaking and that cry louder than the falls for a moment, and then swallowed and he was in the air, turning over and over, moonlight finding the silver-white of his sides and dark green back before he cut into the water, the sound lost in the roar. (33)

The friend has turned into a salmon, climbed the falls, and is on his way up the river to spawn and die. The narrator wonders at the end if he will ever feel "strong enough to eat salmon" again (33). At the beginning of the episode, he says, "Someone must see to it that this story is told: you shouldn't think this man just threw his life away" (29). But we are never exactly sure what sort of a story is being told here. Is this a tall tale, a saint's life, an exemplary life, a metamorphic life, a heroic life? I am not sure it makes much difference. By ordinary American standards, this life is nothing and amounts to nothing. It makes you think of Whitman loafing and inviting his soul and of Thoreau sitting all day on the stoop of his cabin at Walden. But in terms of this book, the story told here is about a man who had an extraordinarily close relationship to nature and somehow has the power, at the end, to change life forms and become part of the ongoing life of the river.

This story is never interpreted for us by Lopez. Nothing in this short and often very strange book is ever interpreted for us. Unlike Thoreau, who tends to interpret everything he sees in nature, Barry Lopez does not offer us an explicit framework of meanings in which to place and understand the events and details of this book. A relentless, systematic approach to this text would destroy it. One must be careful and attentive to hear what it is "saying," to perceive what it is doing. In a sense, we must learn to save these notes from our highly trained analytic and hermeneutic minds by realizing that—here anyway—things most often just are, and that to turn them all into symbols is to trespass on, rather than share in, their being. The next section of the book, "The Shallows," makes this danger quite clear.

"The Shallows" consists almost entirely of detailing and naming what is found in the shallows. It is about how one comes to know that stone is stone, about the value of coming to know this place, any place, in detail and in terms of what it is. The value of such experience and knowledge is entirely terminal. What if we just want to hear the heartbeats of salmon roe (39–40), discover the variety of stones in the shallows (38), see the constellations reflected in chips of obsidian glass (40), know what the willows feel like in the dark (40), or, again, lie still long enough to experience birds alighting nearby to talk to each other? Such experiences and knowledge may lead you to the realization that idly throwing stones into the river may violate both the stones and the river. Eventually it may lead you to the knowledge that everything is sentient and that the idea that only humans feel is a function of human arrogance. The only use to which the "I" or narrator puts his knowledge of nature is self-realization about some of the most fundamental shared experiences of all living creatures: terror, dying, loss of "children," community, hunting, self-discipline, triumph, suffering, joy, and many others. These are not exclusive human experiences; they are shared by all living forms, including stones.

The next section, "The Rapids," is written as if it were recorded as part of a news program. The event that organizes it is classic: two men have foolishly tried to shoot the rapids and have been drowned. "Nature" has "killed" them. One of the voices recorded belongs to a man who has always lived by the river, understands it, and does not blame it, though he lost his wife to it. Most of the rest of the voices belong to people who only respond to the two dead men and perceive the river as an aggressor and killer. One of the voices belongs to the reporter who came up to do a story on the drownings. By the end of the section—which is only four pages long—his attitude toward the river and the deaths has been changed somewhat by the replies and comments of the man who lives by the river and understands it. That is all there is to this section. Once again it has taken some of the most common negative human attitudes toward nature and subjected them to scrutiny. Few ideas have so preoccupied the human mind as the idea of nature as predator, as something wild, uncontrolled, inhuman, threatful, and fearful that must be domesticated, humanized, controlled, eradicated, destroyed, and subjugated. Lopez has studied this idea in some detail in *Of Wolves and Men,* and it is common in human experience—right down to fear of insects. The demonizing of nature is one of the most common themes in the history of human civilization.

The next section, "The Salmon," tells a wonderful story about yet another obsessed character who, to honor the salmon he loves and the renewal they exemplify, uses stones he has gathered from the river to build a huge, forty-eight feet long, nine feet high, stone salmon on a small gravel bar in the middle of the river. It takes him more than four years. When he is done, he has conceived and built the perfect male sockeye, with the irregularities of rut—the hooked jaw and the bright red mantel. Balanced on its belly and with its caudal fin swept to one side, it is poised in an explosive moment. Its eyes are made of hand-polished lapis with pupils of obsidian; the teeth are of white quartz. Its jaws are open so that one has a narrow view down its cavernous dark throat. He finishes his fish in September just as the salmon enter the river and begin their annual migration two hundred miles upstream to their spawning ground where he lives. He waits for them that year with an extraordinary sense of accomplishment, in a state of great serenity. The fish do not appear on schedule, and he begins to worry. One night in October, during a rainstorm, he goes out to see if the salmon have come. He passes his perfect fish and heads down the stream toward the shallows. There he discovers the salmon, by the thousands, as far as he can see, behaving strangely. He realizes after a bit that they are turning around and heading back downstream. They have seen or sensed his huge stone fish upstream and will not go near it. "His guts fell away from his heart" when he realized what was happening and why. He understands "the presumption in his act," its intrusion and trespass; he tries to apologize, touching the fish, but they go anyway.

> He brought his hands to his face for a while, in the passing mist of the rainstorm, he imagined what they [the salmon] would say. That it was the presence of the stone fish that had offended them (he tried to grasp the irreverence of it, how hopelessly presumptuous it must have seemed), that it was an order born out of fear, understood even by salmon, to be discarded as quickly as nightmares so that life could go on. (54).

He thinks of dismantling the fish but finally only removes the obsidian pupils from the eyes so that the fish is and will seem blind (or dead). In later years, we are told, he writes poetry about stonework and butterflies and slowly reclaims his life (54).

The episode with the huge stone salmon-monster is so central to an understanding of *River Notes* that we need to examine it in more detail, especially for its metaphorical content. In his epiphany after the battered, half-dead, migrating salmon have turned back down river, the conceiver and builder of this wondrous forty-eight-foot fish is "overwhelmed with an understanding of the *presumption in his act*, made the more grotesque by its perfection" (54). The "act" of course is the creation of the fish out there on the gravel bank. The perfection of the fish makes "the presumption in his act" more grotesque. We are not told what exactly this is but must intuit it from the rest of the epiphany. The perfection of the stone fish in contrast to the battered, half-live salmon creates the grotesqueness. We ask ourselves what his assumptions were when he conceived of and built this fish, and we wonder why the stone fish "offended" the salmon; why, as he imagines it, they would have, as he should have, found it irreverant and "hopelessly presumptuous"; and finally, why the fish "was an order born out of fear," or something conceived and created out of fear. "Order" refers to the ordering of the stone to create the fish, not to an "order" given to the salmon to turn back. The fish is a human act that intrudes into, or trespasses on, nature.

It also interrupts or disrupts the life cycle of the salmon, just as a man would. The salmon will only spawn where they are born, so in turning them back with the fish, he prevents them from completing their natural and mysterious life cycle, which ends with spawning, dying, and helping to provide the detritus on which the new salmon will depend for food in the early months of their life.

This human act is presumptuous because it oversteps due bounds, and it does so out of ignorance or arrogance or both. The creator of the fish did not really understand what he was doing and could not foresee the consequences of his action. He lacked proper knowledge and experience of the salmon, the river, and nature. He broke the law, not in the legal sense, but in the sense that there are, as the ecologists keep telling us, laws of nature that should not, often must not, be broken. As soon as we recognize this basis of its morality, we come to realize that this is a very moral book. The act of building the stone fish on the gravel bar in the river was irreverent, or the reverse of what the man originally thought and intended, because it was disrespectful of the salmon, again because of ignorance or arrogance or both. Worse, or more seriously, it does not honor or show devotion toward the salmon, but rather prevents them from completing their life cycle.

Four more notes or stories follow "The Salmon": "Hanner's Story," "Dawn," "Upriver," and "Drought." They bring this short book to closure.

Anger in the Water—"Hanner's Story"

Hanner's story is a story he was told by the Quotaka medicine man Elishtanak. He tells it to the narrator, who tells it to us. That is the way stories get passed on and some kind of connection between the past and present is maintained. The story being told here goes way back to a time before there were any human people living in the valley by the river and is about how the bears, salmon, and river worked things out among themselves, arriving at agreements and courtesies they each understood and abided by. The bears and the salmon work out an agreement about how many salmon the bears need and can take during each migration. But they forget to make an agreement with the river because neither thought it was necessary. One fall the river will not let the salmon enter from the ocean. There is a lot of talk between the two, and the river finally lets the salmon enter; but when the salmon get up to where the bears live, the river does strange things and will not let the bears get at the salmon. The river talks to the bears and salmon and says that "there had to be an agreement. No one could just do something, whatever they wanted. You couldn't just take someone for granted" (61–62). So they talk. They all say who they are, what powers they have, and what other agreements they have.

> Everybody said what they needed and what they would give away. Then a very odd thing happened—the river said it loved the salmon. No one had ever said anything like this before. No one had taken the chance. It was an honesty that pleased everyone. It made for a deep agreement among them. (62)

And the agreement remains unchanged.

Hanner's story is set in the context of another story about a kind of utopian community that was formed by the river in the 1840s but suddenly breaks up and disperses one night, leaving no trace of itself. Evidence of some local flooding exists, but nothing else. The narrator is puzzled by what happened to the Sheffield community. Hanner says he does not really know, but "maybe they never made an agreement with the river," not knowing that it was necessary, and so the river flooded them out, quickly, one night, and they left. Here, again, we have an account of the relationship between humans and nature, with a distinction being made this time between the way Indians and most whites understand and relate to nature. And we are told—or shown—again, as we are throughout *River Notes,* that nature is sentient and orderly, and that as human beings, most of us know little to nothing about it, about how it works, and about how to relate to it. Hanner tells the narrator that "you can feel the anger in water behind a dam." That says a lot. Maybe we do not need to say any more here and should simply try to ramify Hanner's story from our own experience and history.

Peace in the River—"Dawn"

Dawn is when the nameless woman in this note rises from her marital bed, dresses, and goes down to sit by the river. Dawn after dawn she does this, naked under her print dress, winter and summer. Sometimes she undresses and washes herself, and sometimes she undresses and enters the river, lying on her belly, facing upstream, into the current, holding onto a maple branch. She imagines herself swimming among salmon. Bits of alder leaf get caught in her pubic hair. She is caressed by the river while her husband lies asleep in the house. It is a kind of strange *aubade,* in reverse because she goes to her lover, the river, each dawn. The river smells like apricots at this hour. The mergansers and salmon are still asleep before dawn; weasels come to the river to drink; twenty-one baby ducks swim by; and an owl feeds overhead. She goes to the river at dawn to "overcome her losses" (67) and enters into a profound, experiential, nondestructive relationship with the river and the life on it, under it, beside it, and above it. Her losses are in her human relationships, presumably with her husband. She opens herself to and becomes an acute and accurate observer of the ecology of the river. She clearly shares being with it and, leaving her husband's bed, enters the river's bed. Her relationship with it is erotic without ever being sexual. Nothing more is made of this relationship, and it would be a mistake to freight it with too much meaning, especially with overt sexual meaning.

Anguish and the Source of Rivers—"Upriver"

The river dies and buries itself in the ocean as *River Notes* begins. We only come to the beginnings of the river as the book ends. They are upriver, above the falls, in an area that is largely unknown. The topographic maps do not tell the truth. At the headwaters, beyond what is shown on the maps, the ravens sit meditating, "and it is from *them* that the river actually flows, for at night they break down and weep; the anguish of creatures, their wailing in desolation, the wrenching anguish of betrayals—this seizes them and passes out of them and in that weeping the river takes its shape" (71). The ravens weep for joy; any act of kindness of which they hear brings up a single tear that runs down their black bills and is absorbed in the trickle of the river taking shape (71).

The Cords That Bind the Earth Together in One Piece: Dancing with the Herons—"Drought"

Drought would cause the death of the river, even if the ravens kept on weeping, and the death of all other life that is dependent on the river. The drought comes. The narrator knows it is coming because the river music has changed and he hears notes that are known to him coming from the river as it drops imperceptibly. He begins a series of actions—

he calls them gestures—in response to the dying of the river, which end in his dance with the herons. He puts stranded fish back into the part of the river that still flows. He retreats into a state of isolation. He abstains from water as much as possible because he knows that such "gestures" are important; he has learned the lesson of the man who built the stone fish. He sleeps by the riverbank regularly. He says simple prayers for the river to express his "camaraderie." He exhorts the river, and "in moments of great depression, of an unfathomable compassion in myself," he makes "the agonized and tentative movements of a dance, like a long-legged bird" (77–78). Everything in, around, and dependent on the river begins to die. One night, the narrator dreams of a certain fish—a fish that has grown so large it is trapped in a single deep pool from which it can never escape. The narrator dreams the fish is dying and sets out, immediately, to find it. The pool in which the fish lived has become a pit, and the fish lies at the bottom of it, dying. Wrapping the fish in his wet shirt, and watering it as best he can, he carries it to what is left of the river and releases it.

This is the paradigmatic act of the book. The act originates in, or is initiated by, a dream, away from the waking cognitive rational centers of the mind. The act puts the narrator in "danger," but as he says, he "knew that without such self-assertion," no act of humility has any meaning. The act is a completely "selfless" act by a human toward a natural creature and, we assume, saves the life of the fish. Afterward, he tries to dance the dance of the long-legged birds again. He "danced it," he says, because he "could think of nothing more beautiful." We are not dealing with the actions of a medicine man or a rainmaker, and this story has little or nothing to do with homeopathic magic. There is no talk here of appeasing angry gods and no identifiable single act has occurred that would bring on the drought, as the acts of Oedipus bring the plague to Thebes. The drought just occurs, as part of a natural rhythm and cycle.

Some humans try to exploit the drought by offering to buy up the property of those who live beside the river; others complain mindlessly about it; and weather forecasters make news out of it. The narrator's reactions and actions must be seen in contrast to these other human responses. Traditionally, humans have personified nature and depicted it as responding to human actions or reflecting suffering and joy; they have viewed nature as something that, in the larger scheme of things, is there to serve human needs; or they have viewed and treated it primarily as an exploitable resource, without much regard for the consequences of their actions. Unlike the bears, the salmon, and the river, they have not been much concerned with "the obligations and the mutual courtesies involved" in the relationship between humans and nature. The narrator, however, like Hanner and Elishtanak, has a very fine and delicate understanding of these matters. He has wonderfully intuitive natural manners. He is tender and compassionate. He is willing to take risks for nature and is not afraid of being foolish.

This last story in *River Notes* tells the story of the whole text better than any other, and as usual, its power is in the telling and in the great precision and imaginative beauty of its details. "The turning point" in the drought comes, we are told, "during the first days of the winter." There is a great gathering of creatures, including the narrator, at what was left of the river. The creatures are all referred to in sentient, dignified terms. Not the lynx but Lynx and Deer and Raccoon and Porcupine come to the gathering. So do Weasel and White-Footed Mouse, Blue Heron and Goshawk, Badger and Mole. Blue Heron speaks:

> We were the first people here. We gave away all the ways of living. Now no one remembers how to live anymore, so the river is drying up. Before we could ask for rain there had to be someone willing to do something completely selfless, with no hope of success. You went after that fish, and then at the end you were trying to dance. A person cannot be afraid of being foolish. For everything, every gesture, is sacred.
>
> Now, stand up and learn this dance. It is going to rain. (80)

Then the dance with the herons begins. We all danced, the narrator says, meaning, presumably, all the creatures. They first dance to the old river songs the narrator remembers from long ago. They dance until the words become incomprehensible and only the music remains; they dance until the music becomes the sound rain makes when it is getting ready to come into a country. Perhaps they dance up the rain. But it seems more likely that the narrator dances back to rain; back to the beginning and source of life itself; back to before words, to before music; back to knowledge of a way of living that the first people (Blue Herons) already knew and practiced; back, finally, to knowledge and experience of the cords that bind the earth together in one piece (81). Everything is sacred. We all share being with each other.

The drought ends. The rains come. *River Notes* ends with the narrator sticking out his hands into the recrudescent river and experiencing the great flow of life and earth-being coming into him from his direct contact with the river. He knows that he too is a part of the cords that bind the earth together in one piece. He understands what it means to be able, to be allowed, and to be asked to dance with herons.

3

Many of the essays in *Crossing Open Ground* are concerned with the breaking of the cords that bind the earth and all its life forms together in one piece and with the consequences of this breaking for humans and nonhumans alike. "Contemporary American culture," Lopez says in "The Passing Wisdom of Birds," "has become a culture that devours the earth" (198). What we need, more than anything else, is "a change in attitude toward nature," a "fundamentally different way of thinking about it than we have previously had, perhaps ever had as human beings" (198). Instead of devouring nature and alienating ourselves from it, we need to figure out "how to get ourselves back *into* it"—by which Lopez means reestablishing the kind of relationship between humans and nature that is depicted in the last part of *River Notes*. "The question before us," he says, is "how do we find a viable natural philosophy, one that places us again within the elements of our natural history" and makes it possible for us to establish and live in accordance with a dignified and honorable relationship with nature (199). There is a "destructive madness that lies at the heart of [our] imperialistic conquest" of nature (197). It is causing us to lose the "focus of our ideals," our sense of "dignity, of compassion," even our sense of mystery (204). This loss, more than anything else, concerns the idealistic, deeply moral Lopez. He seems to believe that there is still hope.

> The philosophy of nature we set aside eight thousand years ago in the Fertile Crescent we can I think, locate again and greatly refine in North America. The New World is a landscape still overwhelming in the vigor of its animals and plants, resonant with mystery. It encourages, still, an enlightened response toward indigenous cultures that differ from our own, whether Aztecan, Lakotan, lupine, avian, or invertebrate. By broadening our sense of the intrinsic worth of

life and by cultivating respect for other ways of moving toward perfection, we may find a sense of resolution we have been looking for, I think, for centuries. (204)

Here is a kind of ultimate statement of what drives or motivates much of what Lopez has written.

Why is nature so central in all of this; why are the wild animals, the wilderness, and the landscape of wilderness so important in Lopez's scheme of things? Because they are the Other that was there before humans (the past), the nonhuman (not the superhuman), the other that is necessary for the dialectic of health and sanity. At the very end of "The Passing Wisdom of Birds," Lopez puts these matters very nicely, discussing the importance of our experience of wild animals and birds.

> The [wild animals] continue to produce for us a sense of the Other: to encounter a truly wild animal on its own ground is to know the defeat of thought, to feel reason overpowered.
> ... It is the birds' independence from the predictable patterns of human design that draws us to them. In the birds' separate but related universe we are able to sense hope for ourselves. Against a background of the familiar, we recognize with astonishment a new pattern. (208)

In this passage, the overwhelming sense of the Other—of the possibility of an order before and apart from human reason, of actions not based on human thought but still as purposeful as any action that is based upon human thought, and of a coherence that is totally independent of human motives and order and values—is so important to Lopez. This loss of self—this being taken out of the self and being able to experience both without a sense of threat, without wanting to destroy the cause of it or demonstrate in some way one's superiority to it—is so essential to getting back into a healthy, sane, and generative relationship with nonhumanized nature. We have to admit its intrinsic worth and its inalienable right to be what it is. We have to learn again that there are other sources of wisdom besides humans. We have to learn what *River Notes* teaches us.

Crossing Open Ground is rich in symbols, symbolic actions, and passages that articulate, in different ways, the message of this book—a book with a mission. At the end of "A Reflection on Snow Geese," Lopez writes:

> We must search in our way of life, I think, for substantially more than economic expansion and continued good hunting. We need to look for a set of relationships similar to the ones Fields admired among the Eskimos. We grasp what is beautiful in a flight of snow geese rising against an overcast sky as easily as we grasp the beauty of a cello suite; and intuit, I believe, that if we allow these things to be destroyed or degraded for economic or frivolous reasons we will become deeply and strangely impoverished. (38)

In his essay on floating down the Colorado River, "Gone Back to Earth," Lopez says: "Each day we are upended, if not by some element of landscape itself then by what landscape does, visibly, to each of us. It has snapped us like fresh-laundered sheets" (43). Later in the same essay, he says, "With the loss of self-consciousness, the landscape opens" (44). And at the end of the essay and trip, he concludes:

> I do not know, really, how we will survive without places like the inner Gorge of the Grand Canyon to visit. Once in a lifetime, even, is enough. To feel the stripping down, an ebb of the press of conventional time, a radical change of proportion, an unspoken respect for others that elicits keen emotional pleasure, a quick, intimate pounding of the heart. (52–53)

Being able to hear your heart beat—a literal and symbolic statement—is what is really important about such an experience. All the value terms Lopez uses here are away from egocentricity and the rational centers of the self.

After arguing—in "Landscape and Narrative"—that the interior landscape (the mind, the sensibility) is deeply affected by the individual self's experience of the exterior landscape, Lopez says, "The exterior landscape is organized according to principles or laws or tendencies beyond human control. It is understood to contain an integrity that is beyond human analysis and unimpeachable" (66). It is the dialectical Other so necessary to what Lopez considers the formation of a healthy self. "Yukon-Charley: The Shape of Wilderness," an account of a canoe trip Lopez took on this Alaskan river, contains many of Lopez's most articulate statements about the importance of wilderness in our lives and about our obligations toward it. We have, he says, "an ethical obligation to provide animals with a place where they are free from the impingements of civilization. And, further, an historical responsibility to preserve the kind of landscapes from which modern man emerged" (81). Lopez argues that "as vital as any single rationale for the preservation of undisturbed landscapes is regard for the profound effect they can have on the direction of human life" (81). Lopez goes on in this essay to enumerate some of these effects, finding among them humility, serenity, and revitalization for someone who has spent too much time in "the highly manipulative, perversely efficient atmosphere of modern life" (82). These are perhaps commonplace ideas about the value of wilderness, as Lopez admits. However, his next point is not so commonplace and is central to all of Lopez's thinking about wilderness. It has to do with "sublime encounter[s] with perhaps the most essential attribute of wilderness—falling into resonance with a system of unmanaged, non-human-centered relationships"—and the ways in which such encounters fulfill deep essential yearnings and needs in human beings (82). Such encounters are not only aesthetic but spiritual and help us to "reestablish a sense of well-being with the earth" (82-83). Lopez speaks often of "the elevation of spirit" that takes place in a wild place, of the quiet rejuvenation that comes to a person there, and of the special kind of "wisdom" one acquires from these experiences (90).

In the essay "Children in the Woods," Lopez describes a scene that might have come right out of *River Notes* and that makes the same point that book does. Lopez and a child are on a mud bar, by a river, and are kneeling by the footprints of a heron, making handprints next to them. Both the child and the man have a sudden revelation about the nature of life itself. It is a striking image, linking animal, human, and place (the river), past and present, old and young. It links many forms of life together and affirms, as ecology and Lopez always do, that "it all fits together" with one form of life affirming another—all others. We are not aliens here; nor is nature our enemy. This powerful sense of the wholeness of all life—of the human and the nonhuman Other— pervades Lopez's writing everywhere. It is, he says, a kind of knowledge from which a "long, fierce peace" derives, a kind of knowledge that "nurtures the heart, that cripples one of the most insidious of human anxieties, the one that says, you do not belong here [on earth], you are unnecessary" (150-51).

Finally, a passage from "Searching for Ancestors," an essay about trying to recover the culture of the Anasazi, puts Lopez's concern with nature in its proper, larger perspective (the moral context of all his work).

> As the Anasazi had a complicated culture, so have we. We are takers of notes, measurers of stone, examiners of fragments in the dust. We search for order in chaos wherever we go. We worry over what is lost. In our best moments we remember to ask ourselves what it is we are doing, whom we are benefiting by these acts. One of the great dreams of man must be to find some place between the extremes of nature and civilization where it is possible to live without regret. (178)

Yes! And to this resonant wish one should add some passages from the profound and eloquent epilogue to *Arctic Dreams*. The extremes of nature and civilization (high technology, trophy and bounty hunting from planes) are presently in mortal combat in such places as the Arctic and the Amazon rain forest, and in lesser places all over the earth. Lopez calls this the "inevitable" "collision of human will with immutable aspects of the natural order" (411). As advances in science and technology (both products of human reason and intrinsic to the evolution of the human species) increase our ability to "alter the land" and to alter (tamper with) the very nature of life itself through gene-splicing, the difficulty of finding someplace—that is, ground—between the extremes of nature and civilization where it is possible to live without regret increases, sometimes exponentially, and a dignified and honorable relationship with nature (the living land) becomes harder and harder to work out and sustain.

The conscious mind is what humans bring into (out of?) nature, and the conscious mind perceives that the human condition is full of paradoxes, ironies, and contradictions—chief among them the collision course described above. One must learn to take responsibility for a life lived in the midst of such paradoxes and contradictions: "There are simply no answers to some of the great pressing questions. You continue to live them out, making your life a worthy expression of leaning into the light" (413). When Lopez bows to the Arctic (414), he seems to have found the ground on which it would be possible to live without regret.

> The conscious desire is to achieve a state, even momentarily, that like light is unbounded, nurturing, suffused with wisdom and creation, a state in which one has absorbed that very darkness which

was before the perpetual sign of defeat. Whatever world that is, it lies far ahead. But its outline, its adumbration, is clear in the landscape, and upon this one can actually hope we will find our way. (414-15)

Addendum

With a serious and thoughtful (and thought-provoking) writer like Barry Lopez, one is always waiting for the next major work to find out where, and in what form, he will go next—especially because he writes so well in all three modes and because he is trying to address what are surely among the most important issues of our time. In the two short books that have been published since *Crossing Open Ground* (1988), Lopez has gone both backward into myth time, in the Indian animal fable *Crow and Weasel* (1990), and forward, in the eloquent essay *The Rediscovery of North America* (1990), into how we can transcend (not undo) our past exploitive and destructive relationship to our native land. Lopez knows that we cannot live in the past as if the present never happened, but he does believe that we can draw on the wisdom of the past to help us find a new (old) direction for the future. He believes that we cannot go on destroying what we found in the New World (already old, with many established native cultures and hundreds of different languages) when we arrived here five hundred years ago from our old world.

Crow and Weasel tells an exemplary tale of a journey of exploration and discovery undertaken by two, young, Northern Plains Indian men who travel far up into the (to them) unknown North, all the way to the tundra, where they meet other native Americans—the Inuit—with whom they exchange gifts and share stories, many of which they find they have in common. At the end of six months, Crow and Weasel return to their own tribe, wiser and better for what they experienced and learned on their journey. The moral—that is, the educational intent—of this simply told and beautifully illustrated fable (which seems to have been written for younger readers) is to exemplify a lost but still viable native wisdom about the biospheric community as a whole (plants, trees, animals, land, other people, and the spirit world) and the proper way for humans to behave as participants in this community.

The Rediscovery of North America is about another journey of exploration and discovery, this time by Europeans, which has resulted, Lopez says, in five hundred years of ravaging, pillaging, exploiting, and destroying of the land and animals, the mineral resources, and the native peoples and cultures in North and South America and everywhere in between. Lopez depicts these actions as both criminal and insane and identifies them throughout with greed, self-interest, and the desire for wealth and status—motives that are still with us in the primary values of the American way of life. Another part, or side, to this eloquent, hopeful essay picks up and reiterates many of the points Lopez made in "The Passing Wisdom of Birds"—but with a major difference. There he was primarily concerned with a much larger whole and with what it means to find, to care for, and to be responsible for a home place—in "Passing Wisdom," North America, but more specifically, our various home places in it.

Where next? What can Lopez say and recommend that we do that will be different and better in this culture of greed and exploitation, which is exemplified for many of us by the stock market and savings and loans scandals, by the relentless exploitation and depletion of our natural resources for economic motives, by unscrupulous developers, by megalomaniacal financial empire builders, and by political leadership that even the party faithful can hardly believe in anymore? Lopez has really taken on the established capitalistic American way of life, its political and economic institutions, free enterprise, market economy, and entrepreneurial values. How and where he will come out of this encounter remains to be seen.

Landscape, wild animals, the passing wisdom of birds, wilderness experiences, and love of place will not save us collectively from the mess we are now in, in the United States and worldwide. Generosity of mind and spirit, courtesy, tolerance, integrity, gentleness, moral fervor, and love of family—all primary values everywhere in Lopez's work—are individual salvation devices, not political, economic, or environmental programs. Both these short books are hopeful, uplifting works that contain specific suggestions about and examples of how we can change ourselves to alter our relationship to the world in which we live and how we can learn to love it rather than destroy it. In a metaphorical passage at the end of *The Rediscovery of North America,* Lopez says that what we "face" is a

> crisis of character. It is not a crisis of policy or of law or of administration. We cannot turn to institutions, to environmental groups,

or to government. If we rise in the night, sleepless, to stand at the ship's rail and gaze at the New World under the setting moon, we know we are thousands of miles from home, and that if we mean to make this a true [new] home, we have a monumental adjustment to make, and only our companions on the ship to look to.

We must turn to each other, and sense that this is possible. (RNA is unpaged)

At the end of *Crow and Weasel,* Weasel says to Crow, as they watch a glorious sunset: "It is good to be alive. To have friends, to have a family, to have children, to live in a particular place. These relationships are sacred" (63). We say yes to these fundamental, primary values and generative attitudes, but we will have to wait for another major book to find out what else Lopez has to say, in what form or mode he will choose to appeal to us, and on what grounds.

NOTE

1. Two other short books by Lopez have been published since 1988: *Crow and Weasel* (1990) in the metaphorical, fictional, narrative mode; and *The Rediscovery of North America* (1990) in the research, historical, personal experience mode.

Peter Matthiessen
1927–

> "Dere are days when I very, very bitter. Cause I wore myself out to get to de place where I de best day is in de main fishery of de island, and now dat fishery don't mean nothin. No, mon, de schooners all gone an de green turtle goin. I got to set back an watch dem ones grow big on de Yankee tourist trade dat would not have amounted to a pile of hen shit in times gone back. I got to swaller dat."
>
> —Captain Raib Avers, in Peter Matthiessen, *Far Tortuga*

Peter Matthiessen was born in New York City in 1927. After high school he joined the U.S. Navy, and he was stationed at Pearl Harbor when the war ended. He graduated from Yale University in 1950 and soon moved to Paris, where he founded *The Paris Review* in 1953. Returning to the States, he worked for several years as a commercial fisherman and as captain of a Long Island charter boat. In the years that followed, Matthiessen maintained an ambitious schedule of wildlife research in Africa, Asia, South America, and the United States. In 1961 he joined the Harvard-Peabody expedition to New Guinea. Other expeditions have taken him to the Indian Ocean, the Himalayas, Australia, and back to Africa. His twenty-one books, including field studies, nonfiction narratives, and fiction, reflect his detailed knowledge of these regions and the careful research that leads to his finished work. Today he is a popular writer (most of his books remain in print) who also receives critical praise and his share of literary prizes. In 1979 he received the National Book Award for *The Snow Leopard*. He has lived in Sagaponack, Long Island, since 1960, and as an indication of the significance of Buddhism in his life, he became a Zen Monk in 1981.

Nature writing and field studies form an important part of Matthiessen's writing career, including *Wildlife in America* (1959), *Shorebirds* (1967), *The Tree Where Man Was Born* (1972), and his most recent book,

African Silences (1991). But he is equally interested in human cultures and professions, especially those that resemble sensitive ecosystems because they are under stress or their way of life has become threatened. Works reflecting this interest include the novels *At Play in the Fields of the Lord* (1965) and *Far Tortuga* (1975) and such nonfiction studies as *Under the Mountain Wall* (1972) and *Men's Lives* (1986). Matthiessen expresses his dual interest and concern this way: "I've always been interested in wildlife and wild places and wild people. I wanted to see the places that are disappearing." He celebrates the virtues of such lost or unknown cultures as he dramatizes and celebrates nonhuman life. Readers of Matthiessen become aware of his tragic fatalism regarding the vulnerability of nature and human cultures caught before the developed world's wheels of "progress."

Matthiessen's Voyages on the River Styx: Deathly Waters, Endangered Peoples

John Cooley

The rainforests, wetlands, and coastal waters of the Americas provide both the setting and the subject matter for a significant body of Peter Matthiessen's work. In this essay I will comment on an archipelago of Matthiessen's marine texts, including *The Cloud Forest* (travel essays, 1961), *At Play in the Fields of the Lord* (novel, 1965), *Far Tortuga* (novel, 1975), *On the River Styx and Other Stories* (1989), and *Killing Mr. Watson* (novel, 1990). These texts not only represent a sizable cross section of Matthiessen's writing during the last thirty years but also map an expansive geography, from the rainforests of the Amazon across the West Indies to the tangled density of the Florida Everglades. They also continue his career-long tendency to look to the often-neglected margins that have become the zones of rapid change and, consequently, of human and environmental degradation. In each of these works, Matthiessen explores the theoretically compatible relationship between human societies and natural systems that is threatened by the impact of human greed in conjunction with technology and rapid development. As an experienced naturalist and ecologist, Matthiessen understands the links, the complex web of connections, between these two groups, and he is aware that the health of the biosphere is essential to the long-term health of and even the survival of humankind. Yet even his bleakest descriptions of the degradation of marginal human communities and threatened ecosystems are leavened with richly evocative reminders of the great beauty, variety, and mystery of the earth and its human inhabitants.

1

Cloud Forest, a first-person narrative of Matthiessen's encounter with South American rainforest and mountain lands, serves as a reminder of how recently and how rapidly the slash-and-burn devastation has assaulted the great rainforests. When Matthiessen made his 1961 voyage by steamer up the Amazon and several of its tributaries, the vast rainforests appeared to be intact, nearly untouched. So undeveloped was the Amazon basin at the time (a region roughly the size of the continental United States) that Matthiessen could write:

> Both the idea and the spectacle are moving—so moving that I could not suppress little squeaks of pure excitement. It is difficult to accept that a wilderness of this dimension still exists. That, despite our airplanes and machines, we cannot really enter it but only skirt its edges. This will change, of course, but it will change slowly; today, should this ship stop at any arbitrary point and send a small boat to the shore, the chances would still be heavily in one's favor that no man, dark or white or copper, has ever stood beneath those trees before. (36)[1]

In some ways this passage echoes the first "voyage" (circa 1885) of young Ike McCaslin from Jefferson, Mississippi, into the "Big Bottom" in Faulkner's Mississippi wilderness classic "The Bear." Faulkner likens Ike's approach to the great yet doomed Yoknapatawapha wilderness to that of a small boat approaching a seemingly impenetrable coast. With childlike exuberance, Matthiessen takes his readers by boat through a prelapsarian, Edenic wilderness, which even in 1960 seemed intact and indestructible. But Matthiessen knows that most of the world's former wilderness regions are in a fallen state and that Faulkner's rhapsody to wilderness initiation ultimately leads to the tragic diminishment witnessed in "Delta Autumn," his coda to "The Bear." Even in the rainforests of the Amazon, such change will occur, Matthiessen speculates, "but it will change slowly." Given the rapid slash-and-burn destruction of Amazon rainforests in the last few decades, we can see that even Matthiessen fails to anticipate the rapidity of change. As he travels by steamer up the Amazon and then the Rio Negro into Peru, Matthiessen observes tiny intermittent clearings where grass-roofed villages punctuate the otherwise ceaseless forest. Despite diminished numbers of crocodiles, river otters, and manatees, the river habitat is, he asserts, virtually intact and seemingly indestructible. To Matthiessen, the visitor peering from the deck of a river steamer, it seems like a great dark wall—"a black misshapen silhouette towering above us on both sides, as if the river were passing down a canyon of the River Styx" (37).

Either directly or by implication, Matthiessen uses the River Styx as a trope in each of the texts being considered here, securing a metaphorical common line between them. The irony in Matthiessen's initial use of this chilling Stygian trope (a river so silent and darkly walled it made him feel as though he was being transported to Hades, the land of the dead) is that it reveals more about the Western traveler's response to the Amazon than it does about the region. By describing the river as black, misshapen, and towering, as if descending a dark tunnel to the underworld, Matthiessen echoes the language of Conrad's "Heart of Darkness" and Vachel Lindsay's poem "The Congo." How can a river ecosystem so intensely full of life be identified with Hades? Is this not primarily the knee-jerk response of the Western visitor prejudiced with frightening jungle stereotypes? Of course, Matthiessen may also be anticipating future species and habitat loss.

Because Matthiessen, a seemingly tireless traveler, writes primarily about remote habitats and peoples, we need to consider, for a moment, the relationship between the traveler and that which he encounters. Claude Levi-Strauss has written at length in *Tristes Tropiques* about the difficulty, if not impossibility, of knowing the "Other." After years of painstaking anthropological work at understanding various Brazilian tribal societies, he concluded it is impossible to understand a primitive society except in contrast to our own. As the anthropologist (like the writer) struggles to understand the people he writes about, he "is still governed by the attitudes he carried with him."[2] We cannot help being the children of our own culture, he asserts, but by struggling to cast off our culture and know another, we finally come to see ourselves and our culture through the other as a mirror. Although Matthiessen is widely recognized for his accuracy in describing remote peoples and species, his mythological reference to the River Styx, though only a single example, seems to confirm Levi-Strauss's assertion that one inevitably carries along the baggage of cultural knowledge in one's travels and into the search for understanding of the Other. Ironically, the Amazonian rainforest contains more living organisms and a greater number of species per acre than any other biome on earth. Contrary to Matthiessen's passing Stygian mood, the "River Styx" on which he voyages passes through one of the lungs and nurseries of the planet.

2

At Play in the Fields of the Lord[3] continues and sharpens Matthiessen's treatment of the Amazonian rainforest. Published four years after *Cloud Forest*, the novel suggests a significantly deeper understanding of the tensions and conflicts of this vast region. The narrative tells of the efforts of four misguided American missionaries to convert and thus save the isolated Nairuna people. Their efforts are complicated by and entangled in the aspirations of a Catholic priest who wants to make "rice Christians" of the Nairuna; of Commandante Guzeman, who wishes to eradicate them; and of mercenary fighters Wolfie and Moon, who have guns for hire and are not particular about their victims. This curious cast suggests the ugly development strategies of imperialism in the Third World; the essential elements of this story are being staged even today in dozens of regions that lie at the precarious edge of "development."

The four missionaries arrive with a fortress mentality, embodied in the cleared field and barbed wire they erect to protect themselves from the subjects of their "play." Their industry, in the face of dismal results, ironically highlights the metaphor evoked in the title. In their missionary work, one counts success in souls, but the developers who follow them count in dollars. Their effort is not play but toil; the greater the "harvest of souls" the greater the glory of God. Their work place is the thick gloom of jungle, not fields of pastoral grain.

One of the missionaries, Hazel Quarier, has nightmares that her God is defecating on her for her work in this forsaken land. In her dream, a hole opens up in the roof of a church and spews slime and excrement on the singing worshipers below. Hazel awakens knowing "that hole was the hole of God." In this perverse scenario, the Christianized pastoral trope evoked in the title will be obtained only if the efforts of the missionaries and forces of development are successful at converting (pacifying) the natives, as well as clearing the jungle and planting fields for play. In an interview, Matthiessen has made clear the implicit view of missionaries in the novel.

It is very hard to argue they've done anything but serious harm. What they really do is lay those people open to the worst abuses of civilization; booze, thieving, jail, corruption of all kinds. And because they're uneducated and unsophisticated they're easy prey to every two bit shyster.[4]

When the four missionaries arrived at the Madre de Dios airstrip, they noticed a small fighter plane, bearing the scars of previous battles, and advertising its owners and purpose: "Wolfie and Moon, Inc. / Small Wars & Demolition" (15). Although the Quarriers and the Huebens imagine their work to be of a holy nature, the text suggests through such ironies as Hazel's nightmare that although their weapons differ, the results of their efforts will be identical. With napalm and strafing, Wolfie and Moon, Inc. could do the job faster, but both teams manage to demolish an integral tribal culture and a piece of the Amazonian rainforest.

In *The Lay of the Land,* a study of gender and attitudes toward nature in America, Annette Kolodny argues that since settlement the American landscape has been repeatedly feminized. This feminization carries with it paternalistic (European) male concepts of penetration, conquest, ownership, and domination of the body of the woman/land. As Kolodny expresses it, before the power of the masculine, the feminine, including nature, "is always both vulnerable and victimized."[5] Matthiessen's Amazonian landscape, hardly an exception to Kolodny's thesis, extends the zone of vulnerability to include tribal and powerless people of both sexes, when confronted with the assault of masculine power directed toward Third World development. Although it may be of debatable validity to extend analyses of American development, such as those of Kolodny and Leo Marx, to other lands, a good case can be made that the forces of change in Matthiessen's writing are primarily those of America and the industrial nations of the West. Matthiessen's victims—sometimes blacks and Indians, sometimes women and tribal people, and sometimes threatened species and habitats—are victimized by the same categories of power, even as the particulars change from text to text. The paradox of the travel writer, especially in a shrinking planet, is that to look with care and sensitivity at other people is also to see ourselves and our actions more clearly than before.

As the missionary quartet soon discover, Commandante Guzman does not recognize national law or international standards of human rights; "he is the law here in the jungle, he takes the place of God himself!" (25). He has engaged in a subversive slave trade for years, selling Indians to wealthy planters. Now he is plotting with Wolfie and Moon to bomb the Nairuna and drive them from the land altogether. When seeking entertainment, he surrounds himself with young half-breed and Indian girls, barely pubescent, who are prepared to offer their bodies for food and drink. One such youngster announces to the missionary men, "Ay am Mercedes, Ay am Vir-geen...you focks Indio girl." Later, during a barroom fight between Moon and the Commandante, she tries to cover her nakedness, "hands pressed like fig leaves to her crotch." This image poignantly establishes the textual connection between the vulnerability of female flesh and the vulnerability of the forest to "small wars & demolitions."

Matthiessen cranks the theme of sexual violation a turn further. The Nairuna men will also be violated by the planned assault: one of the desperadoes boasts that when they get at the men, they will "Blow Their Little Brown Pricks to Kingdom Come" (27). This complex image reflects a desire to emasculate the male and violate the female. The text is abundant with such gender-related double entendres, which serve usually to link the prudish and "innocent" missionaries to the detestable plans of Commandante Guzman and other "development" forces. Tellingly, right after hearing the male emasculation threat quoted above, one of the missionaries admits that once an attack on the Nairuna occurs, "they will be softened up for an outreach of the Word...and this will make our work...a darn sight easier"(28). Violated women and castrated males should make easy targets for the word of God and the laying on of the missionary's hand, offering blessing and healing.

Having sufficiently exposed the practices and implications of the missionary project and the local commandante, Matthiessen shifts to the main focus of his tale, the story of Lewis Merriweather Moon. He is of Cheyenne Indian and white parentage but has lost the wisdom and beliefs of the Old Ways during his years of accommodation to the values of the white world. In accordance with his name (an allusion to Merriwether Lewis of Lewis and Clark), Moon embarks on a reverse voyage and exploration of the tribal Old Ways. During a flight over the Nairuna villages, Moon watches a warrior shoot an arrow of defiance at the plane. The event helps bring Moon to his new senses; he becomes aware that he is more firmly connected to these vulnerable Indians than he ever was to the white world. Moon breaks with Wolfie and flies again over the Nairuna, but this time he parachutes into the village clearing as if out of the morning sun. Not only does he save the Nairuna from immediate attack, but they believe him to be a sky spirit who has come to empower them.

Lewis Moon's escape from the forces of exploitation to "play in the fields of the lord" and of the nearly powerless Nairuna can be fruitfully viewed in a pastoral context. All pastorals focus attention on a natural landscape and find in its seeming simplicity and beauty an agreeable interlude from the pressures and constraints of urban or civilized life. The more thoughtful pastorals have attempted to reveal the complexity of the pastoral or middle landscape, which resides between the city and unmodified, wilderness land. As Leo Marx puts it, "the pastoral figure is called upon to mediate between the two worlds and their conflicts, tensions . . . if not to resolve their differences."[6] The pastoral mode finds expression in many literary forms. In fiction it usually involves a quest by a pastoral figure who has disengaged from an oppressively complex or hierarchical society. Despite numerous variations, the quest cycle typically involves loss, challenge, search, and some sort of renewal or reinvigoration. A final phase of the quest cycle finds its completion in a return to and reengagement with society, yet with an invigorated outlook and ecological understanding that has emerged from the pastoral experience.

Leo Marx has also observed that "the ominous sounds of machines, like the sound of the steamboat bearing down upon the raft or the train breaking in upon the idyll of Walden, reverberate endlessly in our literature."[7] It is difficult to find a more poignant illustration of the "machine in the garden" motif than Matthiessen has assembled here the sudden and thundering presence of an armed fighter plane flying over the villages of a people living with Stone Age technology. Matthiessen's variation on the metaphoric pattern has technology's pilot abandon ship. In a manner reminiscent of Faulkner's Ike McCaslin as he seeks "The Bear," Lewis Moon casts off his watch and compass, the impediments of civilization he must jettison to engage in the pastoral quest. Unlike Ike McCaslin, Moon keeps his revolver, a foreshadowing of the partial failure of his pastoral journey.

For a time it appears that Moon may help the Nairuna and other rainforest tribes band together to form a "new Indian nation, the greatest of all time, greater even than the Iroquois" (217). Through unified strength, they might be able to defend themselves for years. Unfortunately, Moon is unable to jettison the background traits that drive him to lead rather than merely advise the Nairuna. Before long, attempts at creating a nation of tribes disintegrates into factionalism and doubts about Moon's divinity. Worst of all, Moon weakens the people he most wants to help by accidentally infecting them with influenza, a disease against which they have no immunity. Thus, the novel implies that cross-cultural partnership and resistance to destructive "development" is enormously difficult if not impossible.

In the final scenes of *At Play,* Moon is again forced to flee, this time from the Nairuna people whom he loves and who have given him liberation. His flight takes him by canoe down another River Styx, literally dotted with black funeral canoes bearing dead Nairuna (primarily from influenza) to their final resting place. Seeking cover, he hides in the funeral dugout of Boronai, the former chief of the Nairuna, wrapping himself in Boronai's shroud as camouflage. Further down this river of corpses, Moon meets Aeore, the young chief and his former friend, now enemy. He is forced to kill the Nairuna he most admires for his strength and courage as a warrior. Then, as if in penance, he conceals himself in Aeore's canoe and hurls away the revolver, his last tie with his former world. Soon he is overwhelmed by the "stink of putrefaction" and the chill of vultures alighting on Aeore's body. His legs and Aeore's touch, "were one, a single flesh. All the cold and rot and smell were seeping into him" (316). Humbled by the wreckage he has caused, an accidental germ warfare instead of the "small wars & demolition" he rejected, Moon pleads to an empty sky for "forgiveness" and his death.

Matthiessen's use of the Stygian, funeral river to Hades suggests possible parallels between Lewis Moon, name bearer of the female orb, and Persephone, daughter of Demeter, the goddess of earthly fertility. In various versions of the Persephone myth, Hades, the king of both decay and the underworld, abducts and possibly rapes his beautiful niece, imprisoning her in the land of decay until Zeus finally arbitrates an agreement between Hades and Demeter. Desperate for her daughter's return, Demeter withholds fertility from the earth for a year. The acceptable outcome results in and gives mythic explanation for the seasonal cycle of the earth from birth and fruition to decay and winter's apparent death. Speculate for the moment that, as his exploratorial name suggests, Lewis Merriweather Moon is still a rising, rather than setting, moon. If, as critic Richard Patterson believes, Moon will be enabled to return "from the fields of the Lord to the field of human involvement" when he sufficiently knows himself,[8] the Nairuna debacle is a tough testing ground to prepare Moon for future challenges. Such a reading is tempting as the final stage of the pastoral quest, but the text is indeterminate regarding Moon's fate. After weeping for himself, Aeore, "the doomed people of the jungle," and American Indians, Moon completed his penance and "felt himself open like a flower." Matthiessen's final image, of the newly flowering Moon, hints, like the greening Persephone emerging from the land of Hades, at the possibility of life's continuity—perhaps for Moon or perhaps for his child, "New Person," born of his union with Pindi, a Nairuna woman. Matthiessen offers no solutions to the plight of the endangered rainforest and its people, unless it is through the vague hope that Moon and others who have experienced pastoral enlightenment will help reset the balance between nature and civilization.

3

In the mid 1960s, Matthiessen paid three productive visits to Grand Cayman Island, visits that provided raw materials for both a *New Yorker* article ("To The Miskito Bank")[9] and his novel *Far Tortuga*. As the *New Yorker* article reveals, tales about the remarkable, albeit threatened, green turtle (*Cheolonia mydas*) and the traditional wind-powered turtle schooners that ply the dangerous waters of the West Indian turtle banks whetted his appetite for knowledge of this region. By a combination of design and chance, Matthiessen visited the Caymans in time to record the phenomena of a vanishing species and of a declining, once-dependent island economy. As he indicated in the *New Yorker* article, he wanted to learn more about the green turtle, "one of the last great reptile relics from the age of the dinosaurs, and an ocean wanderer whose powers of navigation are more awesome than those of the birds"; and he wanted to sail "on a turtler voyage before it was too late" (126).

Matthiessen's difficulty in booking a voyage on a turtle schooner turned into a lucky break. By the time he was able to sail with Captain Cadian Ebanks of the *Lydia Wilson* (Captain Raib Avers of the *Lilias Eden* in the novel), the masts of the once sleek schooner had been chopped in half and she had been modernized with the insertion of twin diesel engines. The "unreasonable" fee Matthiessen was charged to join the *Wilson* crew proved of good value; his article is filled with the details and grim realities of what he regards as a heroic, noble, and dangerous profession. Although he missed the era of the wind turtlers, he experienced something much more poignant: "the changes in the *Wilson*...were only a small part of the metamorphosis that was coming fast to Grand Cayman" (140).

Matthiessen speaks much more candidly of the transition in the *New Yorker* article than in the novel. In the former, he tells of the traditional importance of fishing and cottage industries to the Cayman economy—professions that were about to experience a rapid decline. Tourism, the prosperous enterprise that has become the mainstay of the Caymans, was virtually nonexistent during his first visit, and "suntan oil was all but unobtainable" (131). Two years later (in the mid-1960s), Matthiessen found flocks of sun worshipers, two supermarkets, a gift shop, a night club, and a car-rental agency. Even during his visits, the tourists were arriving faster than the green turtles were disappearing from the West Indies, a marked change for the cottage economy and culture of the Caymans. "To the Miskito Bank" provides a rare opportunity to read the highly polished travel notes from which a major novel subsequently emerged.

Matthiessen has commented that he wishes to "speak for those who cannot speak for themselves."[10] Although speaking for others is a problematical undertaking, as Levi-Strauss makes clear, in *Far Tortuga* Matthiessen attempts to give voice to the natural world and the Third World.[11] The dialogue in *Far Tortuga* is almost entirely a representation of various British West Indian and Honduran dialects. It is also freed from the usual textual designators ("Captain Avers said," etc.) This dialogue, along with the text's spareness of style and rather Zen-like pen-and-ink representations of the sun, moon, stars, and horizon, creates an effect of vastness and starkness to match the open seas of the Caribbean. Matthiessen also experiments with the use of "white spaces" around the text "to achieve resonance, to make the reader receive things intuitively, hear the silence of the wind, for instance, that is a constant presence in the book."[12] While his subjects are netting the disappearing turtles, Matthiessen catches the voices of the last of the green turtle fishermen in the Caymans and gives presence as well to the turtles, the wind, and the sea. These particular efforts are conspicuously absent in *At Play in the Fields of The Lord,* which does not attempt to speak for the Nairuna and the rainforest.

Captain Raib Avers, like his real-life counterpart, is the last of the old-time captains to ply the waters of the Caribbean turtle cays, and his ship, the *Lilias Eden,* is the last of the wind schooners still in operation. Despite the potential for melodrama, Matthiessen avoids the temptation of a sentimental tale about Avers's last voyage and the end of the green turtle fishery. Like the real Captain Ebanks of the *Lydia Wilson,* Captain Avers has partly converted his *Lilias Eden* to a modern, diesel-powered vessel. Having exhausted his savings in the diesels (of which only one works properly) and a radio, Avers is forced to abandon the conversion. Consequently, there are no running lights, fire extinguishers, or life preservers; the radio only receives messages, and the galley is (appropriately enough) a makeshift chicken coop lashed to the deck. Worst of all, to make space for the new engines, the crew sleeps in a shack secured to the deck in front of the ship's wheel. One of the *Eden* crew expresses the dilemma.

> Now dat is a hell of an arrangement. Dat is a HELL of an arrangement, dat is. De mon at de helm cannot even see where de ship GOIN! On all de boats I ever sailed on, I never see nothin to beat DAT! (27)

Because the helmsman gets nothing more than a grand view of the sleeping quarters, the *Lilias Eden* and her crew are essentially cruising in the dark, both mute and blind.

Moreover, Captain Avers's crew darkly suspect that he burned a family-owned schooner, the *Clarinda,* and used the insurance money to finance the pathetic modernization effort. They also speculate about other clandestine and desperate moneymaking schemes of Captain Avers and his half-brother, Captain Desmond Eden: smuggling Cuban sharkskins and transporting illegal firearms. Captain Avers and his brother represent a culture that talks with endearment about turtles and sailing vessels, calling the turtles "girl" and giving the schooners names like *Clarinda* and *Lilias Eden.* Yet Avers apparently burned *Clarinda* for insurance money and assaults the body of *Lilias Eden* until she is barely seaworthy so that he may harvest more green turtles, and as a consequence, he contributes to their near-extinction. Leo Marx observed in *The Machine and the Garden,* regarding nineteenth-century American clashes of technology and pastoralism:

> The sudden appearance of the machine in the garden is an arresting, endlessly evocative image. It causes the instantaneous clash of opposed states of mind; a strong urge to believe in the rural myth along

with an awareness of industrialization as counterforce to the myth. (229)

But Matthiessen's dominant image yokes the violation of the feminine but with the cataclysmic attempt to graft one technology onto its predecessor. This extension of Marx's now-classic thesis suggests the continuing power of the pastoral to contemporary writers and readers as a compelling mode through which to express in contemporary terms the continuing conflict between technology and pastoral values.

Just as *Far Tortuga* encourages several tropological readings of Captain Avers's last voyage, it also invites, almost insists on, historical and cultural dialogue. With the reader's patience, I will both summarize and expand upon Matthiessen's three-page historical preface to the novel. I will also employ Alvin Toffler's thesis in *The Third Wave* to illustrate how *Far Tortuga* links successive waves of technological change and Third World development to vital changes in the lives of turtles and men.

The green turtle is the only turtle whose meat is a valued commodity. The first Spanish and Dutch sailors to explore the Caribbean reported masses of green turtles in the vicinity of the Cayman Islands. Ferdinand Columbus, son of the famous explorer, noted in his journal of 1503 that he had sighted two low islands so covered with and surrounded by turtles that they looked like little rocks and thus were called *Tortugas*.[13] Not only did they look like turtles, but the Cayman Islands were at the center of the turtle trade for three centuries. A report in 1688 refers to forty sloops from various nations simultaneously gathering green turtles in Cayman waters. By the nineteenth century, the waters around the Caymans were already overfished, and turtle fishing soon shifted to the Miskito Banks off the Honduran and Nicaraguan coast. The largest fleet of turtle boats continued to sail from the Caymans, however, a fleet that numbered twenty-five in 1900 and ten or twelve vessels in 1956.[14] As a base for soup and chowder, the green turtle was so popular in England that the 4,109 turtles exported to Great Britain in 1956 made up a sizable part of the annual Cayminian income.

It is not difficult to identify developments in the fishing industry that contributed to the decline in the turtle catch; the same conditions brought the fictional Captain Raib Avers to desperate measures to catch up. The advent of canning, and later freezing, technology made it possible to slaughter and process greater numbers of turtles in Georgetown, Grand Cayman, and other West Indian ports. The diesel conversions of wind turtle schooners, of which the *Lilias Eden* (and *Lydia Wilson*) must have been the last and least successful, followed the changes in processing. The diesel schooners extended the fishing range and safety of a crew and greatly increased the take of turtles. The Cayman Islands' report for 1960 indicated that "a few remaining schooners" continued to operate from Georgetown; the annual report for 1965 concluded that all turtlers were motorized vessels; and finally the 1975 report, a decade after Matthiessen's visits, failed to list a single turtle vessel. Ironically, by this latter date a fledgling industry, Mariculture Ltd. was making significant advances in the sea-farm hatching and raising of green turtles.[15]

Whether by chance or design, Matthiessen visited Grand Cayman at a turning point in its cultural and economic hisory. A year after his excursion on the converted turtle schooner, the Nicaraguan government extended its territorial waters to two hundred miles and revoked a decade-long turtle fishing agreement with vessels from the British West Indies. A year or two later, Matthiessen would have missed altogether the opportunity to describe this disappearing profession. Presumably, he was aware of the overfishing that signaled the end of the turtle fishery and threatened the survival of the species. The Far Tortuga tragedy was also characteristic of commercial fisheries worldwide during the 1960s and 1970s. Because of advances in technology, fishermen had bigger and faster vessels that could fish larger regions more effectively. According to Lester Brown and the Worldwatch Institute, by 1980 eleven major ocean fisheries "had been depleted to the point of collapse," and one of those was the green turtle fishery.[16]

Captain Avers and his crew know that the turtling profession is nearly played out and that his modernization, if it is to pay for itself, will only hasten the decline of green turtles as he extends his range and season to increase his profits. Avers reveals his awareness of some of the changes at work about him.

> De schooners all gone and de green turtle going. I got to set back and watch dem ones grow big on de Yankee tourist trade dat would not have amounted to a pile of hen shit in times gone back. I got to swaller dat. (255)

Raib Avers represents workers of any profession who are unable or unwilling to adapt to changing conditions; like an endangered species, they too will head for extinction. Among those positioned for survival in this chilling and humbling tale of sea deaths are Speedy, who outlives the sinking of the *Lilias Eden,* and Avers's half-brother, Captain Desmond Eden.

When Matthiessen visited Grand Cayman in the 1960s, he saw an island culture also on the brink of rapid change. At the time there were one or two banks on the island and three hundred beds available in small hotels and a few homes. During that period, the traditional exports of seafood, including turtle and shark, and such cottage industries as thatch-rope manufacturing helped the Cayminian society balance its books. The government sent letters to inquiring tourist agencies warning that the Islands had an inadequate electrical supply, bad roads, and lots of insects.[17] Because the Cayman government could see the impending collapse of its traditional cottage industries and fisheries, it followed the established formula for development in the Caribbean. Within twenty years, the Caymans became the condominium and investment capital of the western Caribbean and one of the world's ranking offshore banking centers. They could boast of 360 bank branches and twenty-five hotels with nearly two thousand beds, almost enough space for the 75,000 tourists who were by then spending $12 million annually. Far from economic independence, their economy was increasingly at the will and whim of transnational business.

Captain Avers indulges in a reverie of past glory. His recollections are of a time when "a captain were one of de island's best, he were not some goddom mongrel fella dat has to hide out down among the cays or day put him in jail!" (60–61). Although he is referring specifically to his half-brother Desmond, Avers also chronicles the decline of his profession both in revenue and prestige. In addition to Desmond's trafficking in illegal goods, he has also "sold out" by taking tourists on cruises about the islands.

> Dem Yankees gone to change de ways of de whole island! Sweet Christ, an honest mon can't hardly find a fish no more along de island, day so many of dem tourist boats foulin de sea! And de mon greasin de skids for dem is nobody else den Desmond Eden! (60)

But the Desmond Edens are not likely to have tapped into the new wealth flowing into the western Caribbean. Although there was a bank for every four islanders on Grand Cayman by the early 1980s (when the population was 15,000), and more telex lines per capita than anywhere else in the world, very little of the money in those banks came from or financed Cayman business ventures.[18] As a result of this kind of development, Canadian, American, and European transnational banks controlled the Grand Cayman economy. This phenomenon is not unique. The major banks of the industrial nations have been moving into the most remote parts of the Third World. They are especially fond of offshore governments that allow anonymous accounts and charge no tax on interest earned. Transnational banks absorb small local banks or drive them to extinction. They significantly influence the economic priorities of a nation or region, and they habitually lend to multinational corporations rather than to the riskier, less stable, local businesses.

Matthiessen's Caribbean writing dramatizes the demise of the turtle industry and anticipates the transition from a colonial to a postcolonial Cayman economy in two decades. But the tableau is larger than the Caribbean, as this sketch of world fisheries and transnational banking suggests. Captain Raib Avers is both a particular individual who dies trying desperately to sail his half-converted schooner through the reefs of Far Tortuga and a complex representation of man caught between "de back times" and, as Speedy puts is, "Modern time, mon" (63ff).

Alvin Toffler's study of the impact of rapid change on developing regions, *The Third Wave,* is particularly pertinent to a discussion of *Far Tortuga.* Matthiessen's "back times" are roughly equivalent to Toffler's "First Wave."[19] This phase is characterized by agriculture or aquaculture, self-made equipment, renewable technology, and a self-supporting society—much as Captain Avers remembers his earlier years with pride and longing. The "Second Wave," the industrial revolution, came relatively late to the Caymans, as it did to many Third World nations. Electrical generating stations, airports and modern docking facilities, automobiles and roads, and factories and refineries accompanied those changes. Within a decade, a "Third Wave," to use Toffler's analogy, hit the Caymans and the larger Caribbean—an electronic, white-collar revolution represented by transnational banking and service industries, such as tourism. In my reading, *Far Tortuga* hints at and anticipates these waves of change that have since swept through the Caribbean. The impact of development sufficiently enlarges and historicizes the scope of this novel, without diminishing the power or poignancy of its narrative.

When Captain Avers fails to find turtles in the usual grounds, he takes a desperate gamble and sails for Far Tortuga. By contrast with Speedy's very real destination ("fifty-five acres, mon, all free and clear" [64]), Avers bases his last efforts on a hunch that some of the green turtles will remain and nest at Far Tortuga rather than swim to the Costa Rican nesting grounds. The Captain's fantasy of a pristine place abundant in migrating and perhaps nesting turtles is quickly dashed when he discovers that his half-brother Desmond has deposited a desperate band of Jamaicans on the island for safe keeping until he can run them ashore, as promised, on "de land of opportunity." Thus, Captain Avers's yearning for an idealized pastoral retreat is defeated by desperate strategies for survival in the Caribbean.

Captain Avers had hoped for a bountiful harvest of turtles easy to catch because of their lethargy just before the females crawl ashore to dig nests and lay their eggs. After overfishing, collecting turtle eggs is the greatest danger to the survival of the green turtle. In addition to scavengering the eggs of the next generation, turtle poachers attack the vulnerable females: "Got calipee poachers, too, y'know. Just grab dat turtle and spin her over and carve dat calipee right off, and leave the rest... with her belly laid wide open to de dogs and birds"(77). In passages recording such actions, violations against nature and against the body of the female intersect. In a novel almost entirely devoted to the last days of the old boys of the turtle trade, the female is marginalized and represented largely by the beleaguered and often-slaughtered female turtles. The few women in the novel serve as objects for sexual gratification. Thus, *Far Tortuga* also incorporates the three Stygian themes (sexual violation, environmental degradation, and death-laden waters leading to Hades) that we have already encountered in *At Play in the Fields of the Lord*.

Far Tortuga presents Captain Avers as representative of all those caught between the waves of social and environmental change. Avers is only less culpable and corrupt by degree than his half-brother Desmond. His modernization of the *Eden* is the work of a man caught by the forces of change, uncertain of a course to chart or of how to use the new technology at his disposal. As Henry David Thoreau pointed out in *Walden,* at the advent of the industrial age, "men become tools of their tools." Captain Raib's move toward modernization, cruising blindly at half-speed into a stormy future, is a succinct image of "blind progress," not only in Third World development, but in the "developed world" as well. The text conveys an emotional fondness for the "old ways" of the Caribbean, but it neither sentimentalizes for long nor gives false promise for the future. Even those who survive the rough and deadly seas of this novel—Desmond, Speedy, and some of the turtles—move in opposite directions. Desmond will doubtless continue as entrepreneur and opportunist, carrying lucrative cargoes of tourists, guns, and illegal aliens to their desired destinations. Speedy, the only character to even approach a pastoral vision in this novel, will return to his family and farm in Honduras.[20]

Far Tortuga contains some of the elements of contemporary pastoralism: turtle fishermen ply the Caribbean shoals for turtles that can outweigh a mature sheep; their rich oral narratives weave a tradition and verbal texture around their work; and Captain Avers entertained a dream of settling at his idealized pastoral retreat, Far Tortuga. But this is a fragmented and ironic pastoral in which the shepherds have lost their way, slaughtering their turtle herds beyond their ability to reproduce. Instead of mediating between nature and urban technology, they have appropriated urban values, with ruinous results. If there is a voice for nature here, it comes from Matthiessen, not his Cayminian turtle fishermen. Yet in *Far Tortuga* there is a larger, deeply ecological sense of the diminutive and dependent relationship of humankind to the biosphere. Even after the loss of the *Eden* and her crew in the death-laden waters of the Caribbean, there is a textual awareness of something more, a universal nature, perhaps, that absorbs and mitigates human defeat. *Far Tortuga's* turtle boat cargo of lost seamen and slaughtered turtles is more Stygian flotsam bound for Hades, where it will join the funeral dugouts of the rainforest peoples. But *Far Tortuga* also dramatizes the insignificance of men—the heroes, the strugglers, and the pillagers alike—and suggests the far greater importance (if not for his characters, then for Matthiessen's readers) of gaining environmental knowledge and of seeking harmony with the vast biosphere.

4

The Florida Everglades are the setting for *Killing Mr. Watson* (the first part of a projected trilogy) and the important title story of *On the River Styx and Other Stories*. Even though the novel is the more recent of the two publications, I will discuss them in their internal chronological order.

Ostensibly, *Mr. Watson* is a turn-of-the-century Everglades frontier western shaped in the chilly atmosphere of a death foretold. Unlikely as he is to stray far from the presence of historical and political forces, Matthiessen comes closer than usual to writing a documentary novel.

The narratives by his fourteen characters create the feel, albeit fictive, of distanced objectivity. Surrounding these documents are narrative confessions, recollections, accusations, and testimonials by thirteen narrators who survived Watson and conveniently left their journals and private diaries for local historians and Matthiessen to interpret. In the "author's note," Matthiessen admits that "the great majority of the episodes are my own creation."[21] Whatever the mix of history and fiction, the narrative voices are as distinctively colloquial as they are spare and direct. Their plainness follows the sea-level flatness of the Everglades, but taken collectively they are as intertwined as mangrove roots, as confusing as a map of the Ten Thousand Islands that border the Everglades.

The Watson story is based on a foundation of factual accounts. Such a man lived in this country between 1892 and 1910, and he was gunned down in a mass execution by a self-deputized posse. Through thirty-six documentary passages, we encounter a legend as confusing as the map of the Everglades (circa 1900) that Matthiessen wisely includes with the text. As with all legendary figures, the tales about Watson are so various it is impossible to accept any one as the most reliable.

Although his neighbors knew Watson was a deadly shot and always armed, none knew whether or not he had committed any of the fifty-seven murders of which he was accused. Mr. E. J. Watson kept his distance from folk, and his formalities too; even his wife referred to him as "Mr. Watson." Yet all who conducted business with Watson found him fair, honest, and prompt in payments. Sammie Hamilton was not the only narrator who said Watson had the best homestead farm between Fort Myers and Key West, a self-sufficient homestead on which he "could make anything grow" (17). One can at times picture Ed Watson as a transitional figure who has outlived his youthful frontier violence and wishes to become nothing more than a good farmer and neighbor. Why, then, did ten of his neighbors kill Watson? The simplest answer is: out of horror and fear after five unanswered murders in the islands, all attributed to Watson, and out of their desire, as family and community folk, to protect their own and have some easy nights' sleep. By implication, Watson's rugged individualism and lawless frontier values were in conflict with the settled, agrarian-pastoral values of his murderers.

Readers familiar with Matthiessen's subjects and interests can hardly rest content with such a solution, even though Watson's lurching life and violent death command the center of this novel. Matthiessen's well-established concerns for civil rights, traditional ways of life, and endangered species find representation along the margins of this Everglades mystery. A decentered reading of the margins—the glades and watery edges—of the text produces a different story, a story of the Indians, the settlers, the blacks and women, and the birds, mammals, and reptiles of the Everglades who struggled to live and survive during the pillaging and clearing of the wilderness, as symbolized by the Watson years.

To establish a marginal reading, let us first consider the Everglades as a singular bioregion. Its four thousand square miles of wetland sits atop a vast limestone-floored basin that tips slightly southward toward the Gulf of Mexico. This enormous marsh or glade is fed with a shallow but steady flow of water from the limestone-floored Lake Okeechobee to the north. A river of grass, perhaps the world's widest and shallowest river, flows through the saw grass glades and palm-cypress hammocks on its way to the Gulf of Mexico. In addition to the flowing Okeechobee waters, the Everglades are drenched with fifty-five inches of rainfall annually. This sea-level glade land is "protected" from the Gulf by a ragged, island-studded coast of mangroves.

Although naturalists, especially ornithologists, discovered the extraordinary riches of the Everglades in the nineteenth century, public appreciation of this unique wetland came about remarkably late. Even though Yosemite National Park received partial protection during the Lincoln administration and Yellowstone, the nation's first official national park, was created in 1872, the Everglades were not so designated and protected until 1947. Today more than 20 percent of the south Florida glades have been drained and converted to agricultural use, controlled with ditches and canals to drain and dry the rich muck lands surrounding a now diminished Lake Okeechobee. True to the dreams of Ed Watson's entrepreneurial associates, this remarkable vegetable kingdom of perhaps 500,000 acres of rich muck land produces two and sometimes three crops of vegetables a year for the east coast market. Due to an accelerated rate of wind and water erosion, it is estimated that 85 percent of glades farmland will become unproductive by the year 2000.[22]

Novelist and essayist Wendell Berry has written succinctly about the "El Dorado" obsession in the conquest of the Americas. As Berry puts it, "The first and greatest American revolution . . . was the coming of people who did *not* look upon the land as a homeland."[23] Berry is not denying that many Europeans came to the Americas to establish permanent agricultural settlements, but he argues that the terms of power so favored the conqueror mentality that even settlers and homesteaders were vulnerable to victimization. Conquerors and settlers have fundamentally opposed views of life and land. The exploiter's goal is money and power: asking of a piece of land, how much it can produce in the shortest possible time. By contrast, the settler, with a belief in nurturing and stewardship, asks how much a piece of land can produce while still maintaining its health and without diminishing its future capacity. Exploiters tend to think in numbers: tons of turtles, pounds of bird plumes, acres of timber. The conquistador mentality favors gold and hidden treasure, but if these are exhausted or unavailable, any other lucrative commodity will do: fur, timber, baser metals, turtles, alligator hides, feathers. Superior technology gives the exploiter or conqueror the stronger hand. In each of Matthiessen's narratives, technological superiority is essential to victimization. Recall the "small wars & demolitions" of Wolfie and Moon and the diesel-converted schooner of Captain Avers and his fellow turtlers, and add to these the deadly draw of Ed Watson, the fastest gun in the Everglades. Despite his likeable qualities (he is a successful farmer and decent family man, with a good sense of humor) Watson is a prototypical illustration of the frontier mentality at work. The cheerful exploiter, he has a gaunt smile for his neighbors, who also know he keeps a revolver warm in the sling below his left armpit. He carves his niche and reputation in one of the last corners of America to resist settlement (community values and rule of law). He will not be fenced in by reason or restraint.

Matthiessen's saga of Ed Watson ("the most celebrated citizen this coast ever produced" [180]) is also the story of the conquest of the Everglades, the slaughter of its wildlife. One of the narrators, Bill House, admits that even before the first white settlers, the Calusa Indians traded some egret plumes and otter pelts for gunpowder and whiskey. But once white settlers entered the trade, "the rookeries over by Lake Okeechobee, they was shot out in four years." And no wonder; their plumes were worth their weight in gold and were sent to an insatiable European market. Watson shot up the best rookeries along with the others, and he took the path of despoliation a turn further by shooting in the vast mangrove rookeries during the breeding season. Bill House describes such a scene.

> When them nestlings gets pretty well pinfeathered and squawking loud ... them parent birds ... are going to come in to tend to their young no matter what, and a man using one of the Flobert rifles ...

can stand there under the trees in a big rookery and pick them birds off fast as he can reload. A broke-up rookery, that ain't a picture you want to think about too much. The pile of carcasses left behind when you strip the plumes... is just pitiful, and it's a piss-poor way to harvest.... A real big rookery... hell, it might take you three four years to clean it out, but after that, them birds is gone for good. (47)

Reports of such shootings made their way north, along with the feathers bound for the millinery shops. Soon such egret hunters as Watson and Le Chevelier were accused of outrageous slaughter by the fledgling Audubon Society. But they were joined in this nursery slaughter by nearly all the males in the Ten Thousand Islands, including "mulatto" Indians, such as Richard Hamilton, and a few blacks, including Bill House. In addition to slaughtering the birds, they waged war on the otters, the manatees, and the alligators. During the drought of 1898, 4,500 alligators were slaughtered in three large gator holes alone. By 1908 the gator and otter trade was also exhausted. As Henry Thompson says it, "Even gators can't stand up to that kind of massacre... Yessir, a lot of God's creation was left dying out there."

Trying to survive along the nearly invisible margins of this arena of white power and conquest are a few "mulatto" Indians and several blacks. The Calusa Indians of Florida had been forced further and further south into the Everglades until they finally disappeared by attrition or intermarriage. Even though Richard Hamilton is the grandson of a Calusa chief, he is relegated to survival on the edge of the white-dominated island culture, in social station way below the whites but above the blacks and tribal Indians. In a white society that has labored to remove or pacify the Indians of the Everglades, there is, nonetheless, an obsession with finding the legendary buried treasure of the Calusa Indians. In one of the many ironies of this novel, the treasure is found accidentally by Bill Collier, a homestead farmer, while "digging garden muck for his tomatoes." Collier barely profits from his hoard, however, and the bone jewelry, wood carvings, masks, and clubs are soon packed off to a private collector from Philadelphia, who doubtless knows their true commercial value. Their lands taken from them and their buried heritage pillaged, an even more valuable gift the Calusa could give goes unheeded, unheard: "If you live Indin way, then you are Indin, color don't matter. It's how you respect the earth, not where you came from" (41).

Watson and the first white "settlers" approached the Everglades with a vengeance, as if they were waging war. The "taming" of the wild was a preamble to the dredging and draining of the Everglades and then their commercial development. "Emperor Watson" had ambitious plans to build a deep water harbor and trading center once the Everglades were gutted and drained. He was often heard saying, "You can't hold a good man down." But Watson was of the old frontier school, compared to the next wave of developers: the slick, well-financed bankers, lawyers, and investors. Take, for example, Hamilton Disston, who acquired four million acres of the northern Everglades for a million dollars. His work crew set out with a massive steam dredge from Fort Myers, working its way up the Calusa Hatchee River toward Lake Okeechobee to lower the waters of the huge lake. Again, Matthiessen reminds us of the inevitable clash of technology and wilderness, which he chronicles so succinctly, as the huge dredge churned up "clouds of smoke and noise that drifted for miles across the shining waters... shifted and resettled the vast muds of the Everglades in a mighty paroxysm of misdirected progress" (111). Meanwhile, other exploiter-developers were building a rail line down the west coast to Fort Myers, bringing electricity to the little city, and constructing a port.

Depiction of women is hardly Matthiessen's forte, but here he moves well beyond the binary opposites (fat/thin, ugly/beautiful, whore/missionary) that characterized the women of *At Play in the Fields of the Lord*. Despite their individual differences, they all orbit around Mr. Watson and the other males in the novel: their comments are on domestic topics and carefully worded. The most revealing female character is Watson's thirteen-year-old daughter, Carrie, who has been told by her father that she will soon marry Walter Langford, the son of a prominent family. In her journal, Carrie writes:

> I am scarcely thirteen—can that be old enough to marry?... A grown man twenty-five years old, nearly twice my age, will sleep in the same bed with Miss Carrie Watson!... What's decent about lying down on top of a young girl and doing ugly things without his clothes on! (122–23)

Elsewhere in her journal, Carrie confirms this impression that she is a curious, astute, and articulate teenager. Yet her father treats her as a commodity to enhance his power base. The same phallocentric urges to power and dominion over nature control the lives and experiences of the women in *Killing Mr. Watson*.

One of Matthiessen's narrators, Bill House, comments, "His death weren't no tragedy!" I suspect even Watson would agree with that conclusion: he had no regrets, no remorse for his actions, which presumably included some grizzly murders. His is the story of a restless, trigger-happy man who demanded too much from his neighbors and from his environment. The patchwork narrative gleaned from the margins of this text tells of the transformation of southern Florida from a paradise for tropical bird and marine life into a land poached, pillaged, and drained of much of its human and biological diversity—all within one generation of American history. The killers of Mr. Watson also participated in this larger crime. Thus Matthiessen constructs an unsettlingly postmodern ending. The vilification of Watson concentrates the crimes that—the text will not let the reader forget—are shared to some degree by all. Instead of one or two "reliable" narrative voices, the fragmented text allows readers to constitute any one voice or a combination of a dozen voices: women as well as men, blacks and Indians as well as whites. The danger in maximizing the freeplay between narratives is that no voice emerges sufficiently to hold Watson and his executioners accountable for their crimes. Consequently, the text remains a matrix of voices, much like the intertwined web of mangrove roots that forms the foundation of the Everglades ecology—leaving it to the reader to constitute his or her own reading of the life of big Ed Watson and the assault on the Florida Everglades.

5

The title story of Matthiessen's *On the River Styx* 24 presents a contemporary glimpse of life in the Everglades. Mr. Ed Watson is just a memory in the Everglades, and one is not likely to see corpses floating on the dark waters of Lost Man River. However, Matthiessen's title reminds us we are still navigating the Stygian waters of Hades' realm. Burkett, a Washington environmental lawyer, and his wife Alice have strayed from their urban world and usual holiday venue to the Ten Thousand Islands of this jagged coast. Unluckily, they have landed at Judge Jim Whidden's flea-bitten mangrove fiefdom to fish for snapper and the elusive robalo. They are drawn to Seminole to experience one of the last frontier towns on the edge of Florida's last wilderness. They should have vacationed in Fort Lauderdale with their friends, but like a growing number of urban Americans, they crave the "primitive," seeking something direct, firsthand, not measured out and processed like the usual tourist package. They encounter a forgotten corner of the world even more disturbing, in its environmental decay and lawless racism, than the packaged gloss they have shunned. Judge Jim, who has inherited Ed Watson's mantle, is *the Man* at Seminole; he owns most of the property, including the Calusa Motel (thoughtfully named after the last Indians in the Everglades), and as self-appointed judge, he administers the unwritten law of this separatist dictatorship, bypassed by civil law and the civil rights movement.

As a characteristic of his fictional strategy, Matthiessen seems to relish thrusting characters and readers into unfamiliar territory. In this case, an "Everglades Paradise" beyond the safe and familiar tourist zone reveals a social and racial nightmare. Not only is the sought-after robalo (commonly called "snook") vanishing from the Florida waters, so is commercial fishing altogether. The cause is a century of pollution, dredging, and overfishing. Midnight drug runs have become the only form of fishing that pays along this coast. Even though Burkett is a government lawyer, he finds himself powerless to insist on federal standards of justice in this treacherous, do-your-own-thing domain. As Burkett's wife, Alice, watches her gaunt, dark-skinned boatman pole their skiff through the mangroves, it is no real surprise that she likens the experience to "traveling on the River Styx."[24] Matthiessen's haunting contemporary story of social and environmental decay is a coda to *Mr. Watson,* much as Faulkner's "Delta Autumn" is to "The Bear." When Ike McCaslin declares, "No wonder the ruined woods I used to know don't cry for retribution! . . . The people who have destroyed it will accomplish its revenge," he could as well be speaking for Matthiessen's ruined landscapes.

6

My reading of these Matthiessen texts has emphasized his treatment of ecological and social issues, which are not only prominent themes in his writing but also significant concerns of his readers. Yet it would be a challenge to argue that these texts lean toward any particular ideology or approach to environmental and social justice issues. Matthiessen's reluctance in these works to undercut or vilify preserves a state of indeterminancy. Not only do his texts frequently look to the margins, those watery edges where the powerless and the forgotten entities struggle to survive, but they attempt to give voice and presence to these entities. Matthiessen deconstructs the frontier heroics of each of his quasi-pastoral protagonists: Lewis Moon, Captain Avers, and Ed Watson. His exploiters are also victims, the hunted as well as the hunters. Of the three, only Moon ("who had never wept in twenty years") seeks forgiveness and catharsis. Only Moon may have achieved the earth wisdom and social wisdom that could earn him the designation of pastoral hero, but Matthiessen does not venture a suggestion as to his fate.

In these texts, Matthiessen favors the season of Hades and almost relishes rubbing his readers in the blood, feathers, and corpses that litter his rivers of Styx. Each of the four examples of Matthiessen's fiction considered here contains recognizable elements of the pastoral, yet each is fragmented and problematic. He is not alone among contemporary nature writers in adopting the tone of pastoral irony and the form of pastoral fragmentation. His writing shows painstaking concern for the state of nature, but it does not overtly mediate, as a traditional pastoral, between nature and civilization. Although Matthiessen's interviews and nonfiction writing reveal great concern over the condition and prospect for humankind and the biosphere, his fiction dramatizes habitat destruction yet shuns any temptation to articulate models of resolution and fruition, or an embracing ecological vision.[25] Levi-Strauss's comment that the traveler trying to find the other finally finds himself bears repeating. Matthiessen gives his reader-travelers ample encouragement to see themselves and to ponder their choices and actions in his and their own desperate landscapes.

NOTES

1. *The Cloud Forest* (New York: Ballantine Books, 1961).
2. Eugenio Donato, "*Tristes Tropiques:* The Endless Journey," *MLN* 81 (1966), 172.
3. *At Play in the Fields of the Lord* (New York: Random House, 1965).
4. From author's personal interview with Peter Matthiessen, August 1990.
5. *The Lay of the Land* (Berkeley: University of California Press, 1979), 143.
6. "Pastoralism in America," in *Ideology in Classic American Literature,* ed. Sacvan Bercovitch and Myra Jehlen (London: Cambridge University Press, 1987), 43.

7. *The Machine in the Garden: Technology and the Pastoral Ideal in America* (New York: Oxford University Press, 1964), 15–16.

8. "*At Play in the Fields of the Lord:* The Imperialist Idea and the Discovery of the Self," *Critique,* 21, no. 2 (1979): 5–14.

9. *New Yorker,* 18 October 1967, 120–27.

10. An informal comment made to the author by Matthiessen, during his visit to Western Michigan University, November 1985.

11. *Far Tortuga* (New York: Bantam Books, 1975). For additional discussions see Richard F. Patterson, "Holistic Vision and Fictional Form in Peter Matthiessen's *Far Tortuga,*" *Rocky Mountain Review* 37, nos. 1–2 (1983): 70–81; and Bert Bender, "*Far Tortuga* and American Sea Fiction Since *Moby Dick,*" *American Literature* 56, no. 2 (1984): 227–48.

12. For this and other remarks by Matthiessen about writing *Far Tortuga,* see George Plimpton, "The Craft of Fiction in *Far Tortuga:* Peter Matthiessen," *Paris Review* 60 (1974): 79–82.

13. For this reference and a full history of the green turtle fishery, see James J. Parsons, *The Green Turtle and Man* (Gainesville, FL: 1962).

14. Ibid., 49.

15. *Cayman Islands: Report for the Years 1959 and 1960* (London, 1961), 15.

16. Lester Brown et al., *State of the World 1985: A Worldwatch Institute Report on Progress Toward a Sustainable Society* (New York: W. W. Norton, 1985), 73.

17. Tom Barry, Beth Wood, and Deb Preusch, *The Other Side of Paradise: Foreign Control in the Caribbean* (New York: 1984).

18. Ibid., 218.

19. *The Third Wave* (New York: Random House, 1980).

20. Although their nesting grounds on the Costa Rican coast are now partially protected, the green turtle remains on the endangered species list.

21. *Killing Mr. Watson* (New York: Random House, 1990).

22. *The New Encyclopedia Britannica,* vol. 6 (Chicago: William Benton, 1978). Ironically, this is the productive muckland depicted in Zora Neale Hurston's novel, *Their Eyes Were Watching God,* which celebrates the tough independence of a black woman.

23. *The Unsettling of America* (San Francisco: Sierra Club Books, 1977; Avon Books, 1978).

24. *On the River Styx and Other Stories* (New York: Random House, 1989), 139.

25. See particularly his nonfiction texts, *The Wind Birds, Wildlife in Africa,* and *African Silences.*

John McPhee
1931–

> On the geologic time scale, a human lifetime is reduced to a brevity that is too inhibiting to think about. The mind blocks the information. Geologists, dealing always with deep time, find that it seeps into their beings and affects them in various ways. They see the unbelievable swiftness with which one evolving species on the earth has learned to reach into the dirt of some tropical island and fling 747s into the sky. They see the thin bank in which are the all but indiscernible stratifications of Cro-Magnon, Moses, Leonardo, and now. Seeing a race unaware of its own instantaneousness in time, they can reel off all the species that have come and gone, with emphasis on those that have specialized themselves to death.
>
> —John McPhee, *Basin and Range*

> This pretty little stream is being disassembled in the name of gold.... Am I disgusted? Manifestly not. Not from here, from now, from this perspective. I am too warmly, too subjectively caught up in what the Gelvins are doing. In the ecomilitia, bust me to private.

In this excerpt from his book *Coming into the Country,* John McPhee wittily demotes himself to a "private" in the ranks of militant environmentalists. Why? Because when McPhee writes, he tries to present all sides, fairly, equally. Whether it is oranges or debates between militant preservationists and resort developers, McPhee lays out an overview, then anchors it in conversation with the men and women who are involved, with the experts—the geologist, the miner, the craftsman—letting them speak for themselves. They are the story, McPhee the recorder. He masters his subject, then presents it in the round.

Writing in *Atlantic,* critic Benjamin Demott observed that McPhee's *Coming into the Country* "manages simultaneously to represent fairly the positions of the parties in conflict—developers, conservationists, renegade individualists" Diane Johnson, in her book *Terrorists and Novelists,* puts it this way, "An outdoorsman, romantic but also astute and accepting, he [McPhee] understands the wilderness, he appreciates naivete, and he also sees who will sell out whom." Some have criticized McPhee's work for its lack of personal material. McPhee dismisses the charge saying, "pointless.... You can't be all things."

McPhee was born in 1931 in Princeton, New Jersey. He received his A.B. at Princeton University in 1953 and pursued graduate work at Cambridge University. Most of McPhee's books came out of work first published in the *New Yorker,* where he works as a staff writer. He also is Ferris Professor of Journalism at Princeton University, where he teaches a course each spring term.

Because McPhee addresses a wide range of subject matter, it could be argued that he writes as a generalist rather than in the nature or pastoral genre. But his respect for the wilderness, his sympathetic portrayal of the independent individualist, and his celebration of the craftsperson refute that contention. Perhaps it might be more accurate to say that when McPhee writes, whether it be about geology or canoes, about environmental conservation or preservation, he does so from his readers' perspectives—acting as a mirror to our independence and loyalties, our prejudices, our avarice, and our humanness. He explores our perspectives, ties them together with his subject in precise, dense prose, and produces a coherent and revealing panorama filled with complexity and frustration—an offering that leaves the task of discovering possible solutions to his readers. As he said to the editor of *Sierra Magazine* (May 1990): "Let the judgements entitled 'The Literature of Fact' fall where they may. Let the words play themselves out."

John McPhee:
The Making of a Meta-Naturalist

Thomas C. Bailey

The essays in this book give evidence that, in an era when nature seems under attack around the globe, we are graced with a sunburst of powerful nature writers, widely varied in interest, tone, subject matter, style, and approach. From Hoagland to Berry, Lopez to Gould, Abbey to Matthiessen, Dillard to Ehrlich, the men and women who are writing today about nature are a compelling bunch, whose writing has urgency and beauty and a broad appeal unusual in the fragmented world of modern reading and publishing. One of the most difficult and prickly of these contemporary thinkers about nature is John McPhee. In his career as a nature writer, he has grown less concerned with nature as pastoral and more and more obsessed with the notion of nature as system, a system inherently at odds with any social or economic system men and women have ever devised—and more gloomily, with any system they ever will devise.

Since long before Virgil, the pastoral has presented nature as a refuge from culture, society, and the noises of war. For the most part, the pastoral has been concerned traditionally with the here and now, with nature as a retreat from a society/culture that in its essence pollutes, spoils, and corrupts as it goes. Conventionally, pastoral nature offers a pure, substantial "here," a free-floating temporal "now" away from the, by definition, frantic concerns with getting and spending, development, and the clash of human wills. The writers of the nineteenth century add a new insight: the pastoral could be used not only to deal with what was wrong with culture but to deal with what was wrong with technology. Thus writers like Thoreau could caution against the machine in the garden in such passages as the following from *Walden:*

> That devilish Iron Horse, whose ear-rending neigh is heard throughout the town, has muddied the Boiling Spring with his foot, and he it is that has browsed off all the woods on Walden shore, that Trojan horse, with a thousand men in his belly, introduced by the mercenary Greeks! Where is the country's champion . . . to meet him at the Deep Cut and thrust an avenging lance between the ribs of the bloated pest? (174)[1]

Thoreau's obsession with the locomotive is not singular, nor is his perception that the Deep Cut through which it runs is a wound in the earth's body. This ambivalence about technology gives the pastoral added appeal to writers of the nineteenth and the twentieth centuries; the waiting myth is simply enlarged, as industrial society is seen to feed on the natural world.

The pastoral, in ancient and contemporary hands, is portrayed as vertical. Instead of presenting time as an ongoing process along a horizontal line that stretches forward into the unknown future and backward into a past that is less and less known the farther away it recedes, until time disappears over the horizon in both directions, most nature writers from Virgil to our contemporaries prefer to think of time as a kind of perpetual present. A surprising number of nature books (beginning with Virgil's *Georgics*) organize themselves around the cycle of the seasons and deal, in terms of time, with just one year. *Walden,* for example, need be about only one year, because at some fundamental level in the natural world, all years are approximately the same, repeating cycles of birth, growth, and death endlessly stacked one on the other. Annie Dillard, Maxine Kumin, Sue Hubbell, David Kline, Aldo Leopold, and many others drive this point home with the ordering of their chapters.

At the opening of his career as a nature writer, McPhee is at ease in these two conventions of his genre. Yet as he follows the implications of the subjects he has chosen, and because of the integrity with which his mind moves, he begins to consider the nature of his claims about nature, about the viability of what for him are received ideas. As he does this, his writing begins to trace a slow pulling away from convention, or at the very least a strong challenging of convention. In his more current writing about nature (the three books written so far in his proposed tetralogy *Annals of a Former World: Basin and Range* [1980], *In Suspect Terrain* [1982], and *Rising from the Plains* [1986]—and the more recent *The Control of Nature* [1989]), he seems no longer so interested in nature as a timeless pastoral or as a peaceful Edenic world despoiled by humans. Rather, he proposes that time properly understood is almost incomprehensibly violent, that nature's system is inherently in conflict with human culture, and that as these systems collide, the puny ones devised by human intelligence are doomed to defeat.

Because most of McPhee's early writing (*A Sense of Where You Are* [1965], *The Headmaster* [1966], and *Levels of the Game* [1969]) concerns heroes, the first of his works that could qualify as environmental writing is *The Pine Barrens* (1968). This book, one of McPhee's most enduring popular works, is about a wilderness in New Jersey (of all places) being relentlessly encroached on by people, roads, businesses, houses, even a proposed jetport for New York City.

As McPhee presents his story, the barrens is a doomed paradise outside of time, populated by such people as Fred Brown and Bill Wasovwich, who love and protect this paradise but sense that it is fated to be overrun by the needs of the culture and the people who surround it. As opposed to outsiders, "the pineys" live in harmony with nature and her cycles.

> The pineys had little fear of their surroundings, from which they drew an adequate living. A yearly cycle evolved that is still practiced, but by no means universally, as it once was. With the first warmth of spring, pineys took their drags... and went into the lowland forests to gather sphagnum moss.... In June and July, when the wild blueberries of the Pine Barrens ripened on the bush, the pineys hung large homemade baskets around their necks.... Cranberries followed blueberries in the cycle of the pines.... In winter the cycle moved on to cordwood and charcoal.... Venison, of course, was available the year round. (67–69)

McPhee's most traditional pastoral, *The Pine Barrens* is lyrical, powerful, probing, and finally, elegiac and sentimental: old ways are best, modern values destroy, and we will not see Fred Brown's like again. Yet McPhee establishes his environmental bona fides with this book and proves that conventions become conventions because they rest on truth: our idea of the world is threatened.

In *The Crofter and the Laird* (1970), McPhee tells again the story of a threatened paradise, an old way of life preserved in "a gigantic block of clear plastic," "the grand anachronism" of a medieval relationship between a group of people paying rent and fealty to a hereditary and absentee laird (35). Their lives are lived on Colonsay, a tiny island in the Atlantic off the west coast of Scotland (original home of Clan Phee), and are as bound to nature's cycles as is imaginable in the twentieth century. Their way of life is doomed. As the Laird, Euan Howard, Fourth Baron Strathcona and Mount Royal, puts it, "It is not easy, or practical, to maintain a paradise" (129). The characters who live on this lovely island are remarkable and compelling, especially such men as Donald Gibbie, McPhee's cousin from whom McPhee rents his house, and Andrew Oronsay. Their lives are unbelievably rich and satisfying, so long as they do not long for the excitements of the mainland. But since the new laird determined that this latterday paradise must "wash its face," the population has fallen to 173 people, and because there is little work for people to do, it will surely keep falling. McPhee loves the island and its characters and simply turns away from speculation about what might happen. In the least satisfying ending of any of his books, McPhee lets *The Crofter and the Laird* dribble away in an almost pointless review of old superstitions and folk myths that no one believes anymore, almost as though he were trusting to the luck his ancestors believed in and his cousins find quaint. But both *The Pine Barrens* and *The Crofter and the Laird* ask disturbing questions as they portray and explore the beauties of the threatened natural world.

McPhee's next book, *Encounters with the Archdruid* (1971), seems in many ways to be a response to the questions he had been asking himself about the natural world and our influence on it, about the surviving paradises and how to protect them. The project began with a phone call to David Brower, president of the Sierra Club and, at that time, the best known conservationist in the United States, perhaps the world. McPhee asked Brower if he could write a book about him, the idea of which would be to pair Brower with a leading developer in each of three sections, to let them talk, discuss, argue, and fight it out, with McPhee on hand to record the results of these "encounters with the archdruid." Not only did Brower agree, but so did Charles Park, "who believes that if copper were to be found under the White House, the White House should be moved" (5); Charles Fraser, who believes "that the use of property ought to be planned," (89) and whose best-known development is Hilton Head, South Carolina; and Floyd E Dominy, Director of the Federal Bureau of Land Reclamation, "the nation's chief dam builder." Dominy says he likes "David Brower, but I don't think he's the sanctified conservationist that so many people think he is. I think he's a selfish preservationist, for the few" (168). These clearly are the makings of a good fight, and the book is compelling reading from start to finish, especially the section devoted to the encounter between Brower and Dominy.

This encounter is "centered on the very definition of conservation. Should it mean preservation of wilderness or wise and varied use of land?" (194). McPhee's widely noted objectivity allows him to present both Dominy and Brower as compelling and powerful men, but Dominy, by his forcefulness and good humor, begins to take the book away from Brower. It is far from McPhee's purpose to side with Dominy; but as he presents him, one of the conventional givens of environmental writing—that developers tend to be scoundrels, especially a dam builder like Dominy—comes into question.

McPhee writes:

> In the view of conservationists, there is something special about dams ... that is disproportionately and metaphysically sinister.... The implications of the dam exceed its true level in the scale of environmental catastrophes. The conservation movement is a mystical and religious force, and possibly the reaction to dams is so violent because rivers are the ultimate metaphors of existence and dams destroy rivers.... "If you are against a dam, you are for a river," [declares Brower]. (193)

To Dominy, a dam represents a chance for a different kind of conservation: sport, water impoundment, and electricity. And as the encounter progresses, Dominy's wit and easy companionship clearly puzzle Brower and some of the other rafters down the Colorado River. He may do bad things, but he clearly cannot be labeled a bad man. Walking behind a barefooted David Brower in a cold, pure mountain stream, Dominy bends and takes a drink, joking, "Now I'll have a drink of water that has washed Dave Brower's feet" (209). He keeps the whole group informed on the beauties of dams, and though his arguments with Brower often reach noisy levels of confrontation and end with him snorting "Nonsense!" and walking away, he never holds a grudge and seems genuinely to appreciate his confrontations with Brower and the others. His self-confidence and energy are captivating to McPhee and to the reader. And he keeps coming back; in one instance, pounded in an argument and doused by the rapids, he waited until he had dried off, then struck "a match and lighted the cigar again" (245).

It would be simplistic to state that in his encounter with Dominy, McPhee all at once becomes a contrarian environmentalist. Brower is also an attractive and heroic person. His love for nature is beyond question, and his endless energies have done much to get the contemporary nature movement up and running; his relationship to mountains and mountain climbing, for example, is particularly compelling.

> I like mountains. I like granite. I particularly like the feel of the Sierra granite. . . . I have an urge to get up on top. I like to get up there and see around. A three-hundred-and sixty degree view is a nice thing to have. (234)

In this encounter, McPhee has allowed himself to strongly present both sides of a very disturbing environmental issue. "See *Desert Solitaire*, page 165!" shouts Dominy to Brower at one point, and when McPhee looks up the reference, he finds: ". . . its angry author—Edward Abbey—imagines 'the loveliest explosion ever seen by man, reducing [the Glen Canyon Dam] to a heap of rubble in the path of the river. The splendid new rapids thus created we will name Floyd E. Dominy Falls'" (215). He has allowed a strong spokesperson for development to make the case for safe development as opposed to none. He has raised an interesting dilemma, because Floyd E. Dominy is more engaging than David Brower. (Like Milton, McPhee struggles with the vitality of Satan.) McPhee, as any great writer will do, works in a genre while at the same time challenging its presuppositions.

As if to explore the human question—the attractiveness of energy, the compelling nature of the human personality, and the appeal of the sometime beauties of technology—McPhee's next two books observe men obsessed with technology. *The Curve of Binding Energy* (1974) deals with the fascinating and brilliant physicist Ted Taylor and his concerns about the wise uses and foolish abuses of nuclear energy; again, McPhee's involvement with a powerful human personality allows him a breadth and generosity of concern that an unrelenting focus on the issues might have prevented. *The Deltoid Pumpkin Seed* (1973) is a hymn to a group of passionate men who, even when no one is looking and no one cares, pour their energies into the quest for better, environmentally cleaner machines. These heroes design a lighter-than-air craft that flies smoothly, quickly, and cleanly through the skies, only to discover that the air industry is committed to heavier-than-air technology: their fine new machine turns out to be a witty, splendid failure. Human culture seems at times either dangerously creative or nonproductively creative, and it is almost always blind to what it has done. But McPhee is working his way slowly toward openly acknowledging the inherent contradictions between our human systems and those larger, slower ones of nature. Issues keep getting harder the more McPhee explores them. Edward Abbey, whose anger is simpler and cleaner, and whose mind, however complex, is untroubled by doubt, can display anger by imagining destruction; but McPhee, as a genuinely thoughtful person, wishes to be as true to the truth as he can be. His paradigms inexorably grow more complicated.

In 1975, McPhee returns to nature writing in *The Survival of the Bark Canoe,* a wonderfully textured tale about Henri Villancourt, an American of French-Canadian extraction in New Hampshire who has taught himself to make birchbark canoes as native Americans made them, refusing electrical or gasoline power of any kind, working in a shop heated by wood and lighted by God. As the book begins, he embodies the myth of the nonpolluting craftsperson keeping alive an ancient art, but as the book progresses and McPhee gets to know him better, it becomes apparent that Henri is morally unimproved by his efforts and his expertise. He is in some ways a villain, a jerk who makes jerky... which turns green. And when his food spoils, he simply eats that of the others without saying please. When the men go to the wilderness itself, it becomes more and more obvious to McPhee and Henri's other companions that they are in the company of a man with few woodsman's skills and absolutely no human skills. His ego is literally as big as all outdoors, and the journey down the Allagash is tough going. (McPhee clearly wants this journey to be a Thoreauvian voyage patterned on those Henry David Thoreau took in the 1840s with his guides; McPhee quotes Thoreau regularly, especially the journals and *The Maine Woods.* He is forced to admit by the end of their trip, however, that perhaps the days of Thoreauvian travel are over, and that there is something inescapably self-conscious about late twentieth-century woodcrafts.)

The text is underwritten by two ongoing mysteries: will they get to the end of the river without murdering Henri? and will they get to see a live moose? Henri survives, even though his comrades grow increasingly surly, and the moose is seen at last.

> In what is now the black of night, we are blinded by the oncoming headlights of a many-ton truck, a logging truck.... On and on the truck approaches, an omnivorous machine, swallowing earth and sky.... Suddenly, though, there is another sound, distinct from the engine's churning. The truck is forcing something up the road, something moving in flight before it, that now, within inches of our windows, pounds by. A hoofbeat clatter, a shape as well... a terrored eye.... A bull moose. (215)

This image of the terrorized moose fulfills many of the conventions of nature writing: the ruination brought by the machine into the garden, the fear natural things have of that machine, and the apocalyptic nature of that confrontation. But in its shapes and explorations, McPhee's book at the same time challenges some of those conventions. The passionate involvement of the self in the environment and in environmental projects might be expected to but does not confer sainthood; Henri Villancourt (as even his name suggests) may make fine canoes, but he makes bad company. Another of the conventions, that "we" are the good guys and "they" are the bad guys, also takes a beating. Passionate environmentalism does not by its nature result in moral soundness. McPhee is concerned about the environment, but he sees a growing problem: human culture (its ability to create as well as destroy) must somehow live in harmony with natural systems, but as Henri shows, often even those who are most obsessively concerned with the relationship between the two systems cannot necessarily deal with and understand them. As McPhee details in his next book, this problem is going to bear some looking into.

Coming into the Country (1977) is no doubt McPhee's most popular and widely read book. It is at once a moving and powerful example of environmentalist prose in the American tradition and a clear-eyed and clearheaded challenge to some of the genre's basic assumptions. This study of Alaska explores the idea of wildness as it exists in uninterrupted cycles of time, the absurdity of governmental responses not only to the idea of wildness but to the idea of wilderness, and finally the way people shape that wildness to their own purposes and in turn are shaped by it.

Part 1, called "The Encircled River," the shortest of the book's three sections, is an extended lyrical hymn to the beauties, complexities, and unimaginable vastness of unbroken wilderness. It deals with a trip by five men down the Salmon River above the Arctic Circle (one source for this section's title; another comes from the sense of the encircled as threatened; another from the relationship between circles and cycles). Including themselves, no more than twelve nonnative persons have ever seen this country; it is wildness indeed. This is nature outside of time, the closed, pristine cycle/circle that endlessly repeats itself and that depends for its survival on its uninterrupted vastness.

The men travel without weapons, without machines, leaving no garbage, no trace that they have been there; and they continually travel aware of the presence or possible presence of the barren ground grizzly bear. The grizzly becomes for McPhee the absolute symbol of the world of wildness, and his fear of the bear finally forces us to speculate that we may try to dominate nature because we fear it.

> The sight of the bear stirred me like nothing else the country could contain. What mattered was not so much the bear himself as what the bear implied. He was the predominant thing in that country, and for him to be in it at all meant that there had to be more country like it in every direction and more of the same kind of country all around that. He implied a world.... There had been a time when his race was everywhere in North America, but it had been hunted down and pushed away in favor of something else. (63)

The bear is predominant in another way as well; the wilderness organizes itself around him. Not only does he "imply a world"; he is at the center of the circle/cycle of natural life in the wildness of the Salmon River.

McPhee makes the circle/cycle the central organizing principle of this part of his book. It begins and ends with McPhee trailing his bandanna in the icy water of the river and returns regularly to the idea that nature's cycle is outside of history, the human idea of time.

> The country is wild to the limits of the term. It would demean such a world to call it pre-Columbian. It is twenty times older than that, having assumed its present form ten thousand years ago, with the melting of the Wisconsin ice. (53)

Each year repeats its age-old pattern without interruption, and the cycle continues in pristine purity. This use of the cycle/circle allows McPhee to organize and imagine that which is otherwise unimaginably vast, inhuman, and ungraspable—Thoreau's "unhansell'd globe." This act of the imagination is necessary because the wilderness is so large that it is shapeless; put another way, wildness is utterly antithetical to humanness, and to imagine the wilderness, to give it comprehensible form in language, defeats it and makes it not wild. Even so, McPhee, with his passionate sense of the bear as participant in endless cycles/circles, allows us, at least briefly, to know and feel the bigness of wildness.

> [The bear] began to move upstream by the edge of the river. Behind his big head his hump projected. His brown fur rippled like a field under wind.... He was romping along at an easy walk. As he came close to us, we drifted toward him.... Instantly, he was motionless and alert, remaining on his four feet and straining his eyes to see.... At last, we arrived in his focus. If we were looking at something we had rarely seen before, God help him so was he. If he was a tenth as awed as I was, he could not have moved a muscle, which he did, now, in a hurry that was not pronounced but nonetheless seemed inappropriate to his status in the situation.... He stopped and faced us again. Then, breaking stems to pieces, he went into the willows. (95)

Alaska is also a place where people live and government governs. Part 2, "What They Were Hunting For," deals with the comic search for a "footprint" in which to locate the new Alaskan capital. Seen not in terms of cycles/circles, which are nature's patterns, but in terms of dialectical oppositions, which are patterns useful to human culture, this section of the book opposes "the Sierra club syndrome" to "the Dallas scenario," the urge of many Alaskans to preserve their environment, and the equally passionate desire of just as many Alaskans to develop it for all it's worth. Consequently, the search for a new capital, made possible by the influx of billions of dollars in oil revenue, focuses these conflicts and allows McPhee to be ironic at the expense of the human foolishness that comes immediately into play when such vast sums of money are there for the taking. His bemusement at the damage such foolishness might do to basic human values is not so ironic.

Just as one cycle/circle leads to another in part 1, one dialectical opposition leads to another in part 2. McPhee also deals with the opposition between the Native Americans and the white men and women who have invaded Alaska and now see it as their own. The natives remain puzzled about the entire notion of property and development and have always been amused by the white name given to their sacred Mount Denali; "Who the Hell was McKinley, anyway?" asks Willie Hensley, the native American chairman of the capitol search committee. McPhee understands the tension between Juneau, repository of "old" Alaskan developmental values, and Anchorage, home of the "new" Alaskan developmental values. Those who support Juneau want things to go on pretty much as they have gone in the past, and if the capital has to be moved, they would just as soon it were moved to Anchorage, "because anyone who has built a city like Anchorage should not be permitted to build one anywhere else" (135). Those who support Anchorage have a wilder, more open view of human possibility along the lines of "you do what you feel like doing. . . . There is a street in Anchorage—a green-lights, red-lights, busy street—that is used by automobiles and airplanes" (134).

Other oppositions abound: between Alaska and the lower Forty-Eight; between the smallness of Alaska's population and its ability to make noise; between the "zoners" and those who demand the free use of land; between northern Alaska and southern; between the people and the workers for the state.

> I know a girl who works in a State Office Building, and she knits a sweater there every month. I hate those bureaucrats with a purple passion—the god-damned parasites, walking around doing nothing, sitting on their butts. I'd like to see them go. I'd like to see mosquitoes eat them. I'd like to see them up in the bush. (166)

Most ironic of all is their desire to put a modern state capital in the wilderness.

This comic odyssey ends more or less without issue. There is no move toward synthesis.

> The committee would finally reduce its list to three sites that would appear on Alaskan ballots, and . . . the emphatic preference of the voters would prove to be Willow. By terms of the initiative, the move was to begin by 1980. Opponents of the move would go on

clinging to the hope that, come groundbreaking day, the money would not be there. (178)

Even in Alaska, land of unparalleled wealth and space, dreams can collide with reality; as of this writing, the capital remains unbuilt, even unbegun.

Government in the wilderness may be an oxymoron, but the people who live in the wild (many of whom came there to escape government and who are often the targets of governmental harassment for living—often extralegally—in that wild) appeal to McPhee, and he spends part 3, "Coming into the Country," dealing with them and their lives in considerable detail. The wonderful world of pristine natural cycles that he creates for us in part 1 becomes wilderness as a stage for human endeavor in part 3. Part 3 has an ordering principle hard to follow, even for readers familiar with McPhee, a writer whose sense of form is intricate and whose sense of organization can often be bewilderingly complex. As he considers in passing such questions as the nature of a "good" government for Alaska, he deals with the history of "coming into the country" and the gold rushes that almost invariably caused that coming. Meanwhile, interspersed throughout this longest part of the book (as long as parts 1 and 2 together) are the stories of past adventures and adventurers, men like Amundsen, the Swedish explorer of the early twentieth century, and Leon Crane, and copilot of a B-24 that crashed in the Alaskan wilderness in World War II, who survived his ordeal by happening on a miner's stocked cabin. In general part 1 is about the natural world, and part 2 is about the political world, and part 3 is about what it takes to create a human culture in that vast and often hostile natural world.

McPhee focuses on the people who make up the culture of modern Alaska and divides them into four main types. First, he considers those men and women like Dick Cook and Donna Kneeland who have moved into the bush to live as much as possible the natural life of the trapper, with the aid of only the most necessary machines. Dick Cook, called by one of his friends "the man of maximum practical application," makes about fifteen hundred dollars a year off the land. He prides himself on not needing modern society and has become, by dint of his fifteen years in the bush, "a sachem figure" to the other river people who are trying to live lives of complete self-sufficiency. Not everyone likes him (including, probably, McPhee, to whom he lectures with some frequency), but everyone respects him, even Donna, who is more like his servant than his mate. Among the river people, "the men seem to be waited upon [by women] to an extent that even our forefathers might not have known" (261). The townspeople seem to have a point when they call the river people romantics.

Second, McPhee deals with the difficult characters in the bush who are there to get its gold, who have been there for a long time, and who use technology when its suits their purposes, but who have not brought with them any urge to maintain such cultural institutions as marriage and community. Joe Vogler typifies these crusty old bachelors who live hermitlike lives, mine for gold, and shoot their mouths off like guns. He came into the country in 1944 and has since mined enough gold to be termed prosperous. Afraid of only two things, snakes and claustrophobia, "he chose to live in a country where he had nothing to fear" (315). But since Alaska became a state, the federal government has been taking an interest in these old placer miners and is trying to control the damage they do to the environment. Joe Vogler does not like this interference and says so. "When the bureaucrats come after me . . . I suggest they wear red coats. They make better targets. In the federal government are the biggest liars in the United States and I hate them with a passion" (317). His language is entrancing: "I'd kill the last pregnant wolf on earth right in front of the President at high noon. . . . I believe in my own kind. I believe in gold. . . . Bulldozers. Cats. Earthmoving equipment" (319). He is old-fashioned, even in Alaska, and when confrontation with the federal agents comes, Joe and his noisy colleagues meekly consent, changing their methods to protect the streams that until now they have ravaged. As McPhee ruefully points out, "his violence was all in his rhetoric" (317).

Third, McPhee carefully scrutinizes the native Alaskan population and their complex responses to the Native Claims Act of 1971. Almost all the natives agree with Michael John David that "You use land. You can't own land," but McPhee succinctly points out the problem that remains.

> As a result of the Native Claims Settlement Act, his sense of land—his people's sense of freedom of the land—after ten thousand years is archaic and obsolete. Michael is a stockholder now. . . . For all its benefits and generous presentations, the bluntest requirement of the . . . Act was that the natives turn white. (393)

McPhee's reaction to the "denaturing" of the Alaskan natives by their confrontations with the modern world goes beyond nostalgia into tragedy:

"They seem to hang suspended between a fast-fading then and a more than alien now" (393).

Fourth, McPhee deals with families who live in communities like Eagle and Central, who live difficult lives because of their surroundings, but who maintain surprisingly bourgeois value systems and support a notion of progress for themselves and their children. The quintessential family of this sort is the Gelvins, and McPhee grows close to these people, especially Ed and Stanley, the father and son, whose relationship, he says, "is as attractive as anything I have seen in Alaska" (430). Both men are superb pilots and superb mechanics who can fix anything and who can take machines apart and put them back together with ease: "Living in the bush, you have to" (291). Stanley is also at home driving heavy equipment, especially Caterpillars, the bigger the better, and this skill leads to the Gelvins's brilliant idea, which leads to their epic adventure, which leads to McPhee's admiration.

The Gelvins's love of and commitment to the country is only exceeded by their desire to discover gold, and in his flying around the country, Ed has discovered a distant valley, surrounded by mountains, where they think there might be a goodly supply of the yellow flakes. The trouble is that because it is in isolated, rugged country, they need to get heavy machinery there to put the mining operation into place. Stanley decides that if they can buy a D-9 bulldozer, the largest Cat, with a fourteen-foot blade, he can drive it there, up creeks and over mountains, in the dead of winter, pulling their supplies behind him on a covered sled. It takes five days, fourteen hours a day, in weather that reaches up to zero degrees for only a few hours. When he gets there, his father flies in an airplane equipped with skis, and they leave the Cat until spring, when they fly back in and begin their mining. These operations consist of dismantling the stream, building dams and sluiceways, and in general defacing the area's natural beauty.

McPhee's reaction to the Gelvins's project is perhaps surprising.

> Am I disgusted? Manifestly not. Not from here, from now, from this perspective. I am too warmly, too subjectively caught up in what the Gelvins are doing. In the ecomilitia, bust me to private.... Whatever they are doing, ... they do for themselves what no one else is here to do for them. Their kind is more endangered every year. Balance that against the nick they are making in this land. Only an easy-going extremist would preserve every bit of the

country. And extremists alone would exploit it all. Everyone else has to think the matter through—choose a point of tolerance, however much that point might tend to one side. For myself, I am closer to the preserving side—that is, the side which would preserve the Gelvins. (430)

This reaction seems to me the crisis point in John McPhee's attitudes toward the natural world. Over the course of many books, and most beautifully in the opening passages of *Coming into the Country*, he has examined and detailed the subtleties, beauties, and complexities of the natural world. And in those same books, he has over and over again returned to the problem of how we, as men and women representing human culture, can live in the world of nature, if not in harmony with its cycles, at least as nondestructively as possible. Yet the human enterprise in Alaska strikes him as so compelling and heroic—*natural*, finally—that he makes a major shift in his point of tolerance and a major break with one of the conventions of nature writing: that human culture is ipso facto dangerously destructive and must be restrained. In coming to an understanding of both human activity and natural cycles, "everyone... has to think the matter through," because there are not any easy answers, any responses that are going to make us all happy. If you restrain the Gelvins, you restrain something as deeply satisfying about humanity (to McPhee) as it is rare: real self-reliance. But when Stanley and Ed and their like are uninterfered with, a pristine natural system that exists in an uninterrupted cycle of generativity stands to be destroyed. "Think the matter through." The honesty and anguish in this statement, chief among a strong book's strong traits, make *Coming into the Country* one of the most compelling and powerful pieces of nature writing in the American canon.

This book raises issues for McPhee that he has been thinking about and trying to resolve ever since. From 1970 to 1977, he published seven books; in the fifteen years since the publication of *Coming into the Country*, he has published another seven, among them three volumes of essays or shorter pieces (*Giving Good Weight* [1979], *Table of Contents* [1983], and *The Control of Nature* [1989]);[2] a short, witty book on the Swiss Army, *La Place de la Concorde Suisse* (1984); a book about sailors and the sea, *Looking for a Ship* (1990); and three books of the proposed tetralogy *Annals of a Former World*: *Basin and Range*, *In Suspect Terrain*, and *Rising from the Plains*. The main reason for this slowing down in what has been a remarkably sustained productiveness is because he has had to work so hard on the *Annals,* what one reviewer has called his "rock revelations."[3]

The *Annals,* densely written, closely reasoned books, ostensibly about geology, seem to me to be about nature seen as completely as we can see it, and they are also McPhee's attempt to resolve the conflicts most powerfully raised in *Coming into the Country.* If time is seen as an unbelievably long system in which mountain ranges are created and worn away, oceans come and go, and entire continents change their location and crash into each other, nature is not a closed, timeless system but a violent system beyond our human capacity to be violent. Our human culture, by implication, can then be seen as more or less harmless in the grand scheme of things. Stanley and Ed Gelvin may destroy the pristine beauty of an unnamed Alaskan stream, but time itself will one day obliterate the Himalayas, is obliterating them now even as it creates them. In the contest between orogeny and erosion, McPhee seems to be implying that the Gelvins are the merest spectators. Consequently, McPhee's tetralogy is not just about "rock revelations" but about the very essence of nature, and it represents McPhee's most challenging thinking about the natural world we live in.

The three books published so far have certain things in common. Each takes place along a stretch of Interstate 80, because as Ken Deffeyes points out in *Basin and Range,*

> It is geologically shrewd. It was the route of animal migrations, and of human history that followed. It avoids melodrama, avoids the Grand Canyons, ... but it would surely be a sound experience of the big picture, of the history, the construction, the components of the continent. And in all likelihood it would display in its roadcuts rock from every epoch and era. (34)

In each of the books, McPhee is in the company of a geologist who is accustomed to working in the field on the one hand, and who has lived an interesting and compelling life on the other, so that the writer may continue to balance the human—Kenneth Deffeyes in *Basin and Range,* Anita Harris in *In Suspect Terrain,* and David Love in *Rising from the Plains*—against the scientific. Each book also has as a subtext a different geological problem. The books (in the same sequence as above) deal with mountain building, with the flaws in current tectonic explanations of the formation of the earth's continents and mountains, with the interrelationships between geology and mineral deposits, and consequently, with the uses of contemporary geological theory in exploration, mining, and development.

The books of the *Annals* also have a common obsession: deep time, the profound changes it can imply, and the strangeness of the ways in which humans measure not only time but space. In *Basin and Range,* McPhee puts it this way: "Numbers do not seem to work well with regard to deep time" (20). In *In Suspect Terrain,* he talks about the same problem: "Human time, regarded in the perspective of geologic time, is much too thin to be discerned—the mark invisible at the end of the ruler" (42). In *Rising from the Plains,* he puts it most interestingly.

> This strip mine ... was a point in the world where geologic time and human time had intersected ... human time, full of beepers and board meetings, sirens and Senate caucuses, all happening in micro-temporal units that physicists call picoseconds; geologic time, with its forty-six hundred million years, delivering a message that living creatures prefer to return unopened to sender. (185)

Because McPhee thinks we find the thought of deep time impossible to think about, being only able to think in terms of generations, and then only two forward and two back, each book is informed with the following insight from *Rising from the Plains,* even though it is not always put so bluntly:

> On a geologic time scale, a human lifetime is reduced to a brevity that is too inhibiting to think about.... Geologists, dealing always with deep time, find that it seeps into their beings and affects them in various ways. They see the unbelievable swiftness with which one evolving species on the earth has learned to reach into the dirt of some tropical island and fling 747s into the sky.... Seeing a race unaware of its own instantaneousness in time, they can reel off all the species that have come and gone, with emphasis on those that have specialized themselves to death. (128)

McPhee finds the human ego amazing. It names things millions and millions of years older than itself, and draws boundaries it thinks are permanent, and it acts as it does, McPhee maintains, because it does not understand the true nature of time or the true nature of nature.

When McPhee finishes the fourth book in this series, it will be about California and its geological complexities. The first three books imply its message with their own: human beings are not privileged, and the culture we have created might well be temporary, a minor disruption in the longer patterns of nature.

McPhee's career as a nature writer first fulfilled and then overturned the conventions of his genre. At this point in his exploration of our place in the natural world, he has returned to a piece of conventional wisdom that we would forget at our peril: nature is more powerful than we are, and much more permanent. The strengths of McPhee's vision are not hard to find: a style that allows him to portray the beauties and realities of the natural world as few writers ever have; a sense of order that allows for striking and powerful comparisons; an ability to juxtapose difficult questions in a way that insures that the reader will become his partner in the making of meaning; and a relentless mind that questions and probes and never settles for the easy answer. His weaknesses are equally obvious: his appetite for technological or professional detail often exceeds that of even a devoted reader; and his fascination with the human personality leads him at times to load the dice against the David Browers of the world and come down on the side of a free enterprise humanism that, despite all its charm, poses grave risks to a healthy biota on a healthy planet. Even so, McPhee's work is largely important. *Meta* connotes transcendence, a moving beyond, above. John McPhee, by going above the limits of genre in his impressive body of work and by thinking with care, passion, and integrity about the creation and life of the natural world over forty-six hundred million years of geologic time, has really made himself into a metanaturalist.

NOTES

1. *Walden* (New York: Perennial Library, Harper and Row, 1987).

2. In *The Control of Nature* (New York: Farrar, Straus and Giroux, 1989), McPhee deals with three occasions when human culture comes into direct conflict with natural systems, and he writes wittily and ironically about the contrary needs of the two systems. In the first essay, he details the struggles of the Army Corps of Engineers with the Mississippi River in the Atchafalaya Basin, and he implies strongly that nature will win. In the second, he details how in Iceland, in February of 1973, a group of brave and industrious men kept a volcano from destroying their town, amusedly pointing out that they could only cause it to go where they wanted

it to by the accidents of topography and human miscalculation; if they "controlled" the flow of lava, it was because of an extraordinary concatenation of chances. In the third essay, he talks about Los Angeles, its recurrent problem of mudflows, and how everyone pays taxes so that the rich may live in the mountains. The book's theme is announced by one character in "Atchafalaya": "Man against nature. That's what life's all about" (20). This is lamentably true, and there is no doubt in McPhee's mind who is going to win. He never lets the reader forget the ambiguities and ambivalences of the title.

3. Joan Hamilton, "An Encounter with John McPhee," *Sierra Magazine*, May 1990. This article is based on one of the few extended and in-depth interviews McPhee has ever given.

Gary Snyder
1930–

> Pressure of sun on the rockslide
> Whirled me in dizzy hop-and-step descent,
> Pool of pebbles buzzed in a Juniper shadow,
> Tiny tongue of a this-year rattlesnake flicked,
> I leaped, laughing for little boulder-color coil—
> Pounded by heat raced down the slabs to the creek
> Deep tumbling under arching walls and stuck
> Whole head and shoulders in the water:
> Stretched full on cobble—ears roaring
> Eyes open aching from the cold and faced a trout.
>
> —Gary Snyder, *Riprap*

Gary Snyder's parents came from pioneering families, a significant background for a poet who has been a lifelong pioneer of backcountry poetic form and explorer of Eastern cultures. Snyder was born in San Francisco, but after two difficult depression years, his family moved to a farm in Washington. Thanks to the farm experience, Snyder learned at an early age the pleasures of a farm environment and the rewards of hard work and a frugal lifestyle. Under the influence of his mother's nightly readings, the young Snyder also developed a love for reading and literature. The family moved to Portland during his high school years, and Snyder diversified his interests and work experiences to include acting, radio broadcasting, and copyboy journalism.

In 1947 Snyder entered Reed College, a stimulating hotbed of influences, including the poetry of Pound, Eliot, Lawrence, and William Carlos Williams, all of which became important influences on his work. His senior thesis at Reed was a study of American Indian mythology, an exploration of subjects that would serve him throughout all of his career. As Snyder has commented, "I mapped out practically all my major interests and I've followed through on them ever since."

After college and a year of graduate study at the University of Indiana, Snyder held a variety of backcountry jobs in the far West, including fire lookout and lumberman. In 1956 he moved to Japan, where he lived for the next twelve years, much of it in Kyoto studying Zen and meditation. Ironically, although Snyder is a major figure among the beat poets, he was in Japan during the major years of the beat movement.

Snyder's first published book, *Riprap* (1959), incorporates philosophies and values he absorbed from Asian cultures. It also questions those aspects of Western civilization that have supported the rise of capitalism to dominance and have condoned the conquest of nature. Snyder also presented a vivid prose account of his experiences in Japan in his collection of prose essays, *Earth House Hold* (1969).

Throughout his career, Snyder has continued to draw creatively from Native American and Eastern philosophy and mythology, while continuing his skirmishing with mainstream American and Western values. He tells his readers, "I wish to bring a voice from the wilderness, my constituency," and he has. His critique of the destructiveness of capitalism is balanced, however, by his clear images of nature's endless surprises and of the rewards of frugality and hard work. A poetry almost always grounded in concrete images and details, it embodies William Carlos Williams's famous edict "No ideas but in things." His poetic technique has also been influenced by the discipline of work, as in his familiar "riprap" imagery of carefully selecting and placing stones on a mountain trail, or as Snyder puts it, "lay down these words / Before your mind like rocks."

Gary Snyder's Descent to Turtle Island: Searching for Fossil Love

Ed Folsom

> I pose you your question:
> Shall you uncover honey/where maggots are?
> I hunt among stones
>
> —Charles Olson, "The Kingfishers"

"Hunting among stones"[1]—the poetics of archeology, the search for lost origins—has been a growing obsession in America's contemporary literature. The concern, more and more, is with *descent,* not the classical/theological tradition of descent to a mythical underworld, but a distinctly American tradition of descent to the *land:* an attempt to get in touch with the continent at some point before the white man began to write his history upon it. This desire marks, in fact, a vital shift in the direction of the American imagination; before the twentieth century, our literature and our energies tended to look west and to the future; now they move *down* and toward the past. Instead of looking west, as Whitman did, to perceive a blank wilderness ("A boundless field to fill!") onto which America, "the greatest poem," could be written ("Our republic is," said Whitman, "really enacting today the grandest arts, poems, etc., by beating up the wilderness into fertile farms"),[2] twentieth-century American poets have been engaging in imaginative descents down through the various layers of what America is and has been, back to the aboriginal land itself.

Among contemporary poets, no one has led us further back than Gary Snyder. His poetic quests thus include the reversal of popular (mis)conceptions of "primitive" people.

> All the evidence we have indicates that imagination, intuition, intellect, wit, decision, speed, skill, was fully developed forty thousand

years ago. In fact, it may be that we were a little smarter forty thousand years ago since brain size has somewhat declined on the average from that high point of Cro-Magnon.

Snyder commented in an interview that "We may be the slight degeneration of what was really a fine form."[3]

Humans of the distant past, then—uncontaminated by progressive society, unaddicted to fossil fuel and the industrial/technological complex that was fed by it—are the ones who hold the wisdom, the techniques and attitudes, that would allow us to live and to continue to live, over long periods, on the earth. "The last eighty years have been like an explosion," says Snyder: "We live in a totally anomalous time." So, to reattach ourselves to sustaining traditions, to proven ways to exist and coexist, we must reject present (aberrant) solutions, and instead descend and dig up "The Old Ways" (the phrase serves as the title of one of Snyder's books of essays). "The cave tradition of painting, which runs from 35,000 to 10,000 years ago, is," Snyder reminds us, "the world's largest single art tradition.... In that perspective, civilization is like a tiny thing that occurs very late."[4]

This urge to get in touch with the virgin soil again, to intimately "know the ground you're on," is not original with Snyder, of course. It is a desire which pervades American literature and is particularly intense in our postfrontier times. Hart Crane, for example, ends the "Quaker Hill" section of *The Bridge* with the command:

> break off,
>   descend—
>   descend—

and he obeys his own directive as he makes a descent to the "tribal morn," to a glimpse of Pocahontas before she encountered a white man, to participation in the natives' regenerative corn dance. In the "Descent" chapter of *In the American Grain,* William Carlos Williams, too, concludes that "it is imperative that we *sink.*" The positive center of Williams's vision of America is Daniel Boone, who is able to make "a descent to the ground of his desire." "Let us dig and we shall see what is turned up," says Williams, as he encourages us "to go out and lift dead Indians tenderly from their graves, to steal from them...some authenticity." And as Theodore Roethke begins his continental journey in "North

American Sequence," he seems to be trying to follow Williams and Crane: "Old men should be explorers? / I'll be an Indian. / Ogalala? Iroquois." But Roethke's attempt at descent is a failure. He is forced to transcend America and what it has come to be rather than descend beneath it to get in touch with what it was and is. So he finally merges with the rose in the sea wind on the Pacific shore and leans west toward the Far East, rooted but flowering above.[5]

Gary Snyder, too, leaned west toward the Far East, leaving the mountain forests of America's northwest coast and actually going to Japan to assimilate oriental ways of perceiving; but, as the structure of *The Back Country* vividly demonstrates, he did come *back;* "I began to feel the need to put my shoulder to the wheel on this continent."[6] His journey, then, took him (as the section titles of *The Back Country* indicate) from the "Far West" to the "Far East," but, firmly and finally, "Back"— east again to America's West. In *Regarding Wave* he reexplored the land, touched it affectionately, but not until *Turtle Island* did he make the radical descent, far below what America is, to what—long ago—it was.

Snyder, in his life and in his poetry, is the culmination of many essential American myths: he recycles us to our origins. He reenergizes, for example, both sides of the Christopher Columbus myth/quest: he has fulfilled Columbus's frustrated search for the mysteries of Cathay, and simultaneously he has sought to regain Columbus's initial vision of the New World, to reenvision and maintain that first glimpse of a green world of possibility, unencumbered by European history. Snyder, then, is the twentieth-century extension and fulfillment of Columbus's dream; he merges East and West by bringing his oriental insight to bear on the wilderness beneath present-day America. Snyder thus refers to himself as a "shaman/healer," for *shaman* is a term that applies both to North American Indian medicine men and to eastern Asian priests, and Snyder emphasizes the similarity in beliefs and attitudes, even in origin, of native Americans and Asians.

> There's something where, say, the American Indians and the Japanese are right on the same spot.... Oh, it's all one teaching. There is an ancient teaching, which we have American Indian expressions of, and Chinese, Tibetan, Japanese, Indian, Buddhist expressions of.[7]

The Asian influence on his work has been the subject of much previous criticism about Snyder.[8] I would like to focus here instead on the other major element in Snyder's work: the element that roots him firmly in the soil of this continent. Snyder, in his journeys to the Far East, learned ways to transcend, but his most recent poetry, and especially *Turtle Island,* demonstrates that his powers of transcendence end up in the service of descent, of digging into this land.

1

In *In the American Grain,* William Carlos Williams writes, "The primitive destiny of the land is obscure, but it has been obscured further by a field of unrelated culture stuccoed upon it that has made that destiny more difficult than ever to determine." Snyder, in *Turtle Island,* agrees: "The 'U.S.A.' and its states and counties are arbitrary and inaccurate impositions on what is really here. . . . The land . . . is also a living being— at another pace." Just as Williams tells us to dig and steal some authenticity from the dead Indians, Snyder knows that "civilization has something to learn from the primitive." "There is," Snyder says,

> something to be learned from the native American people about where we are. It can't be learned from anybody else. We have a western white history of a hundred and fifty years; but when we look at a little bit of American Indian folklore, myth, read a tale, we're catching just the tip of an iceberg of forty or fifty thousand years of human experience, on this continent, in this place.[9]

The image becomes clear; we are part of an immense palimpsest; the United States is but a superficial layer, the most recent (and damaging) inscription over a series of earlier texts. Whitman may have found "the United States themselves" to be "essentially the greatest poem," but post-Whitman poets have found them to be a poem that was violently scratched over earlier poems, imposed on and obliterating native texts, native ways, natives themselves. To descend in the palimpsest, then, to learn what meanings lie preserved in the lower layers, to rediscover ways of existing *on* the land without destroying it—this is the basis for the American poet's desire for descent. But the descents become increasingly difficult, for our layer—the text of the American present—has been a violent inscription, one that continues to tear into the continental tablet itself: "Something is always eating at the American heart like acid,"

Snyder says, "it is the knowledge of what we have done to our continent, and to the American Indian." It is difficult, therefore, to get beneath the text of America, for we have destroyed most signs of the texts that preceded us, destroyed so much of the very page we are written *on*. Intent on designing our present, we have made barren much of our past.[10]

So, to begin, Snyder must find that the Indians are not only buried in the actual ground below us but also buried in our imaginations, in the ground of the American psyche; to use a favorite figure (and euphemism) of Whitman's, the natives were "absorbed" by Americans as America expanded into and assimilated the continent. Often unknowingly, we ingested some of their ways, their names, their beliefs, at the very same time we thought we were ridding ourselves of them. The problem becomes one of how to get *at* them, at their *resources*, how to descend through dead layers to what was here before us. "It takes a great effort of the imagination to enter into that, to draw from it," admits Snyder, "but there is something powerfully there."[11]

Snyder's way is to seek out the "back country," and in doing so he performs a history-defying feat, for he crosses the frontier from civilization into wilderness nearly a century after the frontier and wilderness in America had been pronounced dead; it was in 1893, after all, that Frederick Jackson Turner composed the epitaph for the frontier, informed us that, coast to coast, the continent was now settled. The page of America, he said, was written, filled in. But Snyder defies this closure by descending; if the Indians are no longer to be found across a geographical frontier, Snyder will seek them out across a psychic frontier, make an imaginative descent rather than a physical journey. So he retreats across the (supposedly nonexistent) frontier, away from civilization, and squeezes himself into the last vestiges of Western wilderness, where he writes, for the first time since Turner told us that the frontier was gone, the poetry of the virgin land. "No one since Thoreau," notes Sherman Paul, "has so thoroughly espoused the wild as Gary Snyder.... His is not 'white man's poetry' but the 'Indian's report.'" Snyder would return us to native ways of seeing the land, and he begins by renaming things.

> The first step in real seeing is to throw out a European name and take a creative North American name. And the second step is to erase arbitrary and non-existent political boundaries from your mind and look at what the land really is, with mountain ridges, and rivers and tree-zones, and just keep ... following those implications.[12]

Snyder's major accomplishment, then, is a rediscovery and reaffirmation of wilderness, a clear rejection of Turner's (and America's) closure of the frontier. Snyder announces the opening of the frontier again and attempts to push it eastward, to reverse America's historical process, to urge the wilderness to grow back into civilization, to release the stored energy from layers below us. For perhaps the first time in American poetry, we have a white poet writing from the "savage" side of the frontier (Turner defined the frontier as the "meeting point between savagery and civilization"), a poet looking back over it at his (fellow) Americans, and speaking to them from the perspective of "She" (the virgin land, whom Snyder identifies with Gaia, "the original earth-goddess") and her guardian/protectors, the natives.[13]

But the problem, of course, is how to get there—down, under, back in time. In *Earth House Hold* Snyder searches for models, techniques, ways; he looks to Wovoka's Ghost Dance religion, with its belief that "the Buffalo would rise from the ground, trample the white men to death in their dreams, and all the dead game would return; America would be returned to the Indians." But the Ghost Dance seems lost now, beyond recall, so he considers peyote (and other drug) cults that might allow illusory but illuminating descents through the palimpsest to a regenerative past: "Peyote and acid have a curious way of tuning some people in to the local soil ... the human history of Indians on this continent. Older powers become evident." Yet he finds that even drugs are not needed for descents to the past.

> For many, the invisible presence of the Indian, and the heart-breaking beauty of America work without fasting or herbs. We make these contacts simply by walking the Sierra or Mohave, learning the old edibles, singing and watching.

And finally Snyder discovers the way back is through poetry, through imaginative descent, for only in the imagination can he effectively tap the lost resources, the stored imagination of those who were here before us. Snyder knows that the Indian is buried somewhere in our psyche, still breathing; the "American Indian is the vengeful ghost lurking in the back of the troubled American mind," he says, and he goes on to recognize that the American wilderness has its counterpart in our imaginations.

To transcend the ego is to go beyond society as well. "Beyond" there lies, inwardly, the unconscious. Outwardly, the equivalent of the unconscious is the wilderness: both of these terms meet, one step even farther on, as *one*.

His hope is that a new tribe of "White Indians" (the successful descenders; those who have gone down, under, back, and who have united with the ghost remnants they [re]discovered) can resurrect the native ways: "When this has happened, citizens of the USA will at last begin to be Americans, truly at home on the continent, in love with their land."[14] "The Way West," as the title of the second poem in *Turtle Island* tells us, is now "Underground"; below the new violent country is the old way, "Painted in caves, / Underground" (5). Preserved, it waits for our descent to it.

2

Snyder's stance: West, near the ocean; in the past, near the Indian and the wilderness "She." There he takes his stand. He (re)named that place, finally, "Turtle Island": "the old/new name for the continent, based on many creation myths of the people who have been living here for millennia, and reapplied by some of them to 'North America' in recent years" (xi). *Turtle Island* is Snyder's true poem of America, but it is written from the perspective of a *continent* (Turtle Island, the virgin She) that rests uneasily beneath the imposition of America on top of it.

> North America, Turtle Island, taken by invaders
>   who wage war around the world.
> May ants, may abalone, otters, wolves and elk
> Rise! and pull away their giving
>   from the robot nations.
>
>   (48)

The operative word is "Rise!" While Williams and Crane implored Americans to "descend" to their prehistory, to the wilderness and its natives that America had covered over, Snyder implores the wilderness and its natives to *rise,* to take possession again. Once again, Snyder speaks from the far side of the frontier; he is there, across the line, recruiting the remnants of the wilderness for an ascent through the American palimpsest, toward a new birth of Turtle Island ("the old/new name").

When asked if "the poet is essentially a pioneer," Snyder was quick to cut himself off from America's westward movement: "No, I wouldn't say a pioneer. A pioneer clear-cuts an ecosystem and sets the succession phase back to zero again." One thinks of Whitman's "Pioneers! O Pioneers!" with their "sharp-edged axes": "We primeval forests felling, / ... we the virgin soil upheaving." Snyder is, instead, an antipioneer, trying to undo or go beneath the pioneers' work, discarding the dangerous elements of the American past, recycling the good elements: "Poets are more like mushrooms, or fungus—they can digest the symbol detritus." This is the poet's role: to live off the cultural detritus and rise again.[15]

But Snyder's rising, unlike Roethke's, stops short of transcendence; while Roethke retreats from industrial sounds to "The Far Field," escaping the encroaching mower to find his rose in the sea wind, Snyder faces technology and stands firm.

> A bulldozer grinding and slobbering
> Sideslipping and belching on top of
> The skinned-up bodies of still-live bushes
> . . . .
> Behind is a forest that goes to the Arctic
> And a desert that still belongs to the Piute
> And here we must draw
> Our line.
>
> (18)

The title of the poem is "Front Lines," and the battle imagery is appropriate; there is anger in these poems. Snyder is the protecting voice of the continental She, who, because of mistreatment by Euro-Americans, is in danger of death; he descends to her and speaks her language: "She is sacred territory. To hear her voice is to give up the European word 'America' and accept the new-old name for the continent, 'Turtle Island'" (104).

"She" is Turtle Island, and she has been raped and ravaged by the American "He"; his appendages still push into her last wilderness and continue the assault.

> Sunday the 4-wheel jeep of the
> Realty Company brings in

> Landseekers, lookers, they say
> To the land,
> Spread your legs.
>
> (18)

Later in *Turtle Island* Snyder adds resonance to this image by tying it to the Eastern mythology he acquired during his passage of India. After parodying a Whitman catalog by listing all the nonnatural things that now "flood over us," the chemically created continental garbage ("Aluminum beer cans, plastic spoons, / plywood veneer, PVC pipe, vinyl seat covers" that "don't exactly burn, don't quite rot"), Snyder warns that we are at the "end of days," in the "Kali-yuga," the Hindu age of degeneration where men value what is degraded, consume voraciously, huddle in cities. American technology becomes Kali herself, the black anti–Earth Mother who gluttonously devours everything, the negative shadow of the wilderness She. While Whitman, in his persona as the American He, had ecstatically "plunged his seminal muscle" into the continent, now, for Snyder in the twentieth century, the destructive rape seems in its last, grotesque stages: "Kali dances on the dead stiff cock" (67).[16]

The American He has aged quickly and obscenely. Frederick Jackson Turner had imaged the westward movement of America as an awakening body, the East pumping its healthy blood (the blood of civilization) to the West through an expanding circulatory system (roads, railroads, and rivers; lines of commerce). Snyder perceives America still pushing its blood westward, but the health is long gone: "Every pulse of the rot at the heart / In the sick fat veins of Amerika / Pushes the edge up closer" (18).

That "edge" is the frontier, the meeting point of He and She. But the American He, Snyder indicates, carries disease; worse, he *is* disease, a cancer eating away at the land. The American cancer is a central image in the book: "The Edge of the cancer / Swells against the hill" (18). The hill is the breast of She (in Native American mythology, as in other mythologies, hills are often perceived to be the breasts of the Earth Mother); the wilderness She, in other words, has breast cancer.[17] America is the very disease that destroys its own source of nurture; like a cancer, it will destroy the body it lives in, then die for lack of sustenance. In "Plain Talk" Snyder makes the image unmistakable: "The cancer is eating away at the breast of Mother Earth in the form of strip mining" (104). "Fed by fossil fuel," says Snyder, western man's "religio-economic view has become a cancer: uncontrollable growth" (103). The imagery of this continent as the bountiful breasts of a loving mother, an image that extends from Native American mythology to F. Scott Fitzgerald's "fresh, green breast of the new world,"[18] reaches a terrifying culmination with the image in *Turtle Island* of a continental breast cancer. It is clear, then, why Snyder urges us to learn to be "in the service / of the wilderness / of life / of death / of the Mother's breasts!" (77). We must learn "how to live on the continent as though our children, and on down, for many ages, will still be here.... Loving and protecting this soil, these trees, these wolves" (105). We must descend, touch the submerged and suffering She, and learn to love her gently, to protect her, to nurse her as she has nursed us, to join her as "natives of Turtle Island" (105).

Civilization, after all, may well be "on the verge of post-civilization," says Snyder, and so the descent to find how the "primitives" lived is an urgent recrossing of a frontier already declared vanished, a crossing over the "line ... drawn between primitive peoples and civilized peoples" so we can "take account of the primitive world view which has ... tried to open and keep open lines of communication with the forces of nature" (107).

Snyder's work is the logical progression of Williams's search to "rename the things seen, now lost in chaos of borrowed title"; Snyder seeks the true thing lost under a borrowed name. "It could be beautiful, Cincinnati could," Snyder once said; but "to get to know Cincinnati ... means, first of all, you have to get rid of the name *Cincinnati* ... because after all it's the Ohio River Valley, really, that you're looking at. And *Ohio* means *beautiful* in Shawnee. And there you go, you start going back ... "[19]

3

The Process: Go back, get under, dig, descend: through the palimpsest of this continent with its accreted layers of varied culture, beneath the artificial, superficial, forcefully imposed top layer, "The 'U.S.A.' and its states and counties," which are "inaccurate impositions on what is really here" (xi). So *Turtle Island* opens with Snyder chanting himself down through the palimpsest, away from the present, into the dim reaches of prehistory on the continent:

> Anasazi
> Anasazi
>
> tucked up in clefts on the cliffs...
>
> (3)

He descends further than any other poet—further than Williams with his glimpses of the virgin land just before Red Eric's people and Columbus touched it; further than Crane to Pocahontas; further certainly than Roethke to the Iroquois. Snyder often judges poets by the depths of their descent: "Pound was never able to get back to—you know, he could get back to the Early Bronze Age and his imagination couldn't go back any further than that. Olson at least gets to the Neolithic." To get back that far is the essential imaginative act for poets: "I mean their imagination [must be] able to encompass it,... [and] feel comradeship in connection with it."[20]

And so Snyder invokes the Anasazi—the aboriginal, pretribal natives, the cliff dwellers—and evokes the sensation of being one of them: "the smell of bats / the flavor of sandstone / grit on the tongue." Indeed, the poem imitates, in the visual form of its lines, a cave painting—the cliff with its indented clefts, from which hangs the thin word-ladder, with the "trickling streams in hidden canyons" in a long, flowing line at the bottom.

Snyder enters his book at a genesis point far before any other American poet. And toward the end of *Turtle Island,* in "What Happened Here Before," he descends even further, digging far into the prehuman past of the continent: back three hundred million years, he sees "First a sea: soft sands, muds, and marls"; he then rises to eighty million years ago, when there were "warm quiet centuries of rain," and three million years in the past, when the river gorges were cut, and

> Ponderosa pine, manzanita, black oak, mountain yew.
>   deer, coyote, bluejay, gray squirrel,
> ground squirrel, fox, blacktail hare,
>   ringtail, bobcat, bear,
>   all came to live here.

Forty thousand years ago came "human people" and the first "songs and stories in the smoky dark." On this immense palimpsest, a mere 125 years ago, suddenly "came the white man: tossed up trees and /

boulders with big hoses, / going after that old gravel and the gold." And arriving back at the present—"now"—from a journey through eons of time, the poet and his sons sit in the last vestige of the backcountry wilderness, and "my sons ask, who are we?" But their questions are drowned out by "military jets head[ing] northeast, roaring, every dawn." The poem ends, ominously yet hopefully, with a bluejay screeching, "WE SHALL SEE / WHO KNOWS / HOW TO BE" (78-81). The bluejay, native of Turtle Island for thousands of years, *knows* how to survive, while the recently arrived white man flies his machines of destruction, quickly burning up the last of the fossil fuels that have come to sustain his life. Like W. S. Merwin in *The Lice,* Snyder approaches a vision of a world where man is gone, but where life goes on; where man, especially technological man, is seen finally as a short but dangerous aberration in the vast, layered history of earth.

Images of the brutal imposition of white man's culture on the continental palimpsest—a culture that, for whatever enriching layers it added, also tore into, mutilated, earlier layers and the very page itself—are ubiquitous in Snyder's book: "The dead by the side of the road" are native animal-people, each a potential "pouch for magic tools," lying now "all stiff and dry," run over by "trucks run on fossil fuel" which traverse not "our ancient sisters' trails" but rather the new imposed concrete layer, "the roads . . . laid across [which] kill them" (8). Snyder occasionally emerges from the past into and onto the layer of the American present. He can even enjoy it now and then, as in "I Went Into the Maverick Bar": "America—your stupidity. / I could almost love you again." But he always redescends, leaving the freeway of the present.

> We left—onto the freeway shoulders—
>   under the tough old stars—
> In the shadow of bluffs
>   I came back to myself
> To the real work.
>
> (9)

Back in the shadows of the bluffs (cliffs) of Turtle Island, he escapes the bluffs (false, alluring fronts) of American culture, sees through them to his real self, his true nativity.

We see some of America's false fronts—the neon steak houses—in

"Steak," where Snyder plays on the growing separation between our food and ourselves. Lulled by "a smiling disney cow on the sign," we eat the "bloody sliced muscle" served us, forgetting that we are what we eat, and what we eat are overweight cows, stupid and "bred heavy" (in contrast to the vanished strong and active buffalo); on the "ripped-off land" grain is now grown to stuff the "long-lashed, slowly thinking cow" that becomes us (10).

So Snyder digs, down and away from all this, sinks deep, retreats far back in time to a rooted place where the changing cultures of man are but momentary fleetings in earth time.

> To be in
> to the land
> where the croppt-out rock
> can hardly see
> the swiftly passing trees...

Where he descends to, deep in the land, back to Turtle Island, he can say, "I hear no news"; it is "like no Spaniard ever came." He is back and down, in touch with the continent before any nonnatives touched it, and he can be sustained by the same energies that trees are sustained by. As the trees find "tiny sources" of light even in the darkest nights, so Snyder, marooned in the twentieth century, can still gather energy from the "tiny sources" of the past that are, through imagination, retrievable, recoverable—not lost, even if dimmed.

> the tree leaves catch
> some extra tiny source
> all the wide night
>
> Up here
> out back
> drink deep
> that black light.
>   (26)

Snyder agrees with Williams that Americans need to find the true ground of their being: "I would like to see *people* 'grounded.' ... Get the people grounded and the poetry'll take care of itself."[21]

4

The vision: Imposing itself over the natural wilderness, America may at first have appeared victorious, but now it begins to fade, to lose power, and Turtle Island, magically rising through the palimpsest, appears again in the place of fading America.

> The USA slowly lost its mandate
> in the middle and later twentieth century
> it never gave the mountains and rivers,
>   trees and animals
>   a vote.
> all the people turned away from it
>   myths die; even continents are impermanent
>
> Turtle Island returned.

In contrast to Merwin, who envisions America's (and man's) future as dead, crumbling into silence, Snyder's new natives of Turtle Island "look to the future with pleasure / we need no fossil fuel / get power within / grow strong on less" (77).

Is this vision practical, usable, or only a romance, a sentimental fiction? Charles Altieri writes: "Snyder's images of an ideal society remain mere fictions that do not address themselves to the immediate and varied problems of our society.... Conscience doth not make woodsmen of us all." Alan Williamson writes: "It is a dark saying, but in some ways I have more hope for Snyder as a source of cultural continuity and human worth after the world has been changed against its will, than as a voice persuasive enough to prevent disaster." Snyder himself admits: "One is not going to quit just because one knows one's going to lose. And we don't even *know* we're going to lose"; "We've just got to stay on the raft and go through the rapids anyhow." So the vision is a fiction, a psychic model, a structure and set of values that will not change the world, but will change *minds,* will readjust mind-sets. Snyder's vision "speaks for a revolution in awareness," Sherman Paul says; "It is to repossess a new world by dispossessing ourselves of an old dream." The sense in *Turtle Island* is of reality becoming nightmare, of recaptured dream becoming reality—if only in the mind, if only to spur a first subtle shift in attitude, in stance, to help us in "riding the times through."[22]

Snyder, seeking the revolution in awareness, advocates the destruction of America—"I won't let him live. The 'American' / I'll destroy"— and, like an Indian warrior, he will make murderous raids on the white man within himself: "As I kill the white man / the 'American' / in me / And bring out the ghost dance: / To bring back America, the grass and the streams." He even writes a new Pledge of Allegiance for the continent rising from beneath America.

> I pledge allegiance to
>   the soil
> of Turtle Island
> and to the beings who
>   Hereon dwell
>   one ecosystem
>   in diversity
>   under the sun
> with joyful interpenetration for all.

And in *Turtle Island* he occasionally shares Roethke's longing (in "North American Sequence") for the "blast of dynamite" that would rid the rubbish from the streams, or the apocalypse of Merwin's "The Last One," where the American creation is systematically obliterated. But Snyder knows that nature is tough, life is obstinate, and things that have existed for eighty million years can survive conflagrations that will destroy the superficial layer of white culture on this continent.

> Fire is an old story.
> I would like,
> with a sense of helpful order,
> with respect for laws
> of nature,
> to help my land
> with a burn. a hot clean
> burn.
>   (manzanita seeds will only open after
>   a fire passes over.)
>   (19)[23]

The first section of *Turtle Island* is entitled "Manzanita," and it consists of poems for and about the white Indians and other avatars of She who will open up and thrive after the superimposed American culture on the continent is destroyed. The seeds that open only after conflagration are among Snyder's most hopeful images; the lodgepole pine is celebrated in *Myths & Texts* for the same power.

> the wonderful reproductive
> power of this species...
> is dependent upon
> the ability of the closed cones to endure a fire
> which kills the tree without injuring its seed.
> After fire, the cones open and shed their seeds
> on the bared ground and a new growth springs up.[24]

Life will endure through massive destruction; the current manifestations of life can be destroyed, but the potentiality of new manifestations will remain. Ideas, too, will endure the demise of civilization, their seeds protected in the unconscious.

> The traditional cultures are in any case doomed, and rather than cling to their good aspects hopelessly it should be remembered that whatever is or ever was in any other culture can be reconstructed from the unconscious, through meditation. In fact, it is my own view that the coming revolution will close the circle and link us in many ways with the most creative aspects of our archaic past.[25]

Much of *Turtle Island*, then, is a preparing of seeds: guides to what mushrooms to eat, how to prepare a bird for eating—"Taste all, and hand the knowledge down" (51). Snyder echoes Thoreau's advice to simplify, simplify: "stay together / learn the flowers / go light." Those who become tough, suggests Snyder, will, like the manzanita seed, endure and thus preserve possibilities for postcivilization: "In the next century / or the one beyond that, / they say, / are valleys, pastures, / we can meet there in peace / if we make it" (86).

So while Snyder seeks to protect what is left of She—"Virgin / ... many- / Breasted" (82)—he also seeks to store energy, to "get power within / grow strong on less" (77), to put resources into the lower strata of the wilderness within. He seeks a "fossil love," a deep affection that he can draw on, use wisely, and live by: "that deep-buried sweetness brought to conscious thought" (114). The American has dug up the resources of this continent, laid waste to the stored energy of eons of time; he is "addicted to heavy energy use, great gulps and injections of fossil fuel." And the Americans, like heroin addicts, will do anything for their fix: "As fossil-fuel reserves go down, they will take dangerous gambles with the future health of the biosphere... to keep up their habit" (103). And fossil fuel is the stored energy of ancient sunlight; it is not replaced except by slow geological change over countless centuries. "Our primary source of food is the sun," Snyder reminds us (31), and he refers to fuel, food, and thought; we transfer the sun's energy from one part of nature to another by eating, burning fuel, talking, writing poetry.

In other words, Snyder seeks to store the energy, the food, of his native thought, his oriental thought, his Indian religion, in a place where it is accessible to all: that place is *Turtle Island*, the poem. These poems are the new fossil fuel—"fossil love"—being stored for future use, "For the Children." While the American digs up and destroys the earth's fossil fuel, Snyder works to store fossil love, to replant lost ideas, to restore lost energies. This "power within" is an inexhaustible resource, and, says Snyder,

> the more you give, the more you have to give—[it] will still be our source when coal and oil are long gone, and atoms are left to spin in peace. (114)

Those are the final words of *Turtle Island*, concluding the section called "Plain Talk"—angry bursts of explanatory prose, baldly stating the ideas contained in the poetry. Like Whitman in *Democratic Vistas*, Snyder angrily turns to prose when he feels that few will listen to poetry. But it is in poetry where we can "stop and think. draw on the mind's / stored richness" (84). And, warns Snyder, we must replenish those stores of wilderness in our minds, prepare for descents to "the silence / of nature / within" (6).

Turner's westering America was drawn into the wilderness across ever-shifting frontiers by what Turner called "love of wilderness freedom." Snyder's twentieth-century counterpart to this process is a descent into the mind, to the past, to the wilderness freedom that was lost under America, the arduous lover; he is drawn there by "fossil love," an affection that pulls him to the lowest layers of the continental palimpsest as he becomes (finds in his own mind) the stored energy of the native continent and captures that stored energy in his poem of Turtle Island. The poem embodies the poet, who becomes the place, all uniting in their devotion to the archetypal American person: the virgin She, continental avatar of Gaia, the cosmological She. It is fitting that Snyder should follow *Turtle Island* with *Songs for Gaia*, new poems addressed to "Mother Gaia" and "Father Sun"; these poems are intent on literally putting man in his place.

> As the crickets' soft autumn hum
>   is to man,
> so is man, to the trees.[26]

NOTES

1. Charles Olson, *The Distances* (New York: Grove Press, 1960), 11.

2. Walt Whitman, *Prose Works 1892*, 2 vols., ed. Floyd Stovall (New York: New York University Press, 1964), 2:404, 434, 369.

3. Gary Snyder, *The Old Ways* (San Francisco: City Lights, 1977), 16; Peter Barry Chowka, "The Original Mind of Gary Snyder," *East West Journal* 7 (June 1977): 35. Chowka's three-part interview with Snyder appears in the June, July, and August 1977, issues of *East West Journal*. Reprinted in Gary Snyder, *The Real Work*, ed. William Scott McLean (New York: New Directions, 1980), 92–137.

4. Chowka, "Original Mind," June, 36, 35.

5. Ibid., p. 37; Brom Weber, ed., *The Complete Poems and Selected Letters and Prose of Hart Crane* (Garden City: Anchor, 1966), 106; William Carlos Williams, *In the American Grain* (1925; reprinted New York: New Directions, 1956), 214, 136, 196, 74; Theodore Roethke, *The Collected Poems of Theodore Roethke* (Garden City: Doubleday, 1966), 189.

6. Chowka, "Original Mind," June, 29.

7. Chowka, "Original Mind," August, 19, and see also 30; Paul Geneson, "An Interview with Gary Snyder," *Ohio Review* 18, no. 3 (1977): 84; reprinted in Snyder, *Real Work*, 55–82. Cf. Abraham Rothberg, "A Passage to More than India: The Poetry of Gary Snyder," *Southwest Review* 61 (1976): 26–38; Rothberg sees Snyder and his compatriots taking "that next—and giant step from California, Oregon, and Washington across the ocean,... which is both an affirmation of American 'manifest destiny' and a simultaneous denial of it" (29).

8. See especially Bob Steuding, *Gary Snyder* (Boston: Twayne, 1976).

9. Williams, *Grain*, 212; Snyder, "Introductory Note," *Turtle Island* (New York: New Directions, 1974), xi (subsequent references to *Turtle Island* are in parentheses); Snyder, *Earth House Hold* (New York: New Directions, 1969), 120; Snyder, *Old Ways*, 80.

10. Whitman, *Prose*, 2.434; Snyder, *Earth*, 119.

11. See, for example, Whitman's "Song of the Redwood Tree," where the natural wonders of the continent are "absorb'd, assimilated" into the "superber race" of the white man; Harold W. Blodgett and Sculley Bradley, ed., *Leaves of Grass* (New York: New York University Press, 1965), 207; Snyder, *Old Ways*, p. 50. On the relation of Snyder and Whitman, cf. Rothberg's article cited in n. 7 and Robert Kern, "Recipes, Catalogues, Open Form Poetics: Gary Snyder's Archetypal Voice," *Contemporary Literature* 18 (1977): 173-97; Kern compares and contrasts Snyder's and Whitman's open forms. See also Snyder's comments on Whitman in Geneson, "Interview," 94-96.

12. Frederick Jackson Turner, "The Significance of the Frontier in American History," *Annual Report of the American Historical Society for the Year 1893*, 199-227; Sherman Paul, "From Lookout to Ashram: The Way of Gary Snyder," *Iowa Review* 1 (1970): 76, 80; Geneson, "Interview," 87.

13. Snyder sees the closing of the frontier as the major problem in America: "I'll say this real clearly, because it seems it has to be said over and over again. There is no place to flee in the U.S. There is no 'country' that you can go and lay back in.... The surveyors are there with their orange plastic tape, the bulldozers are down the road warming up their engines, the real estate developers have got it all on the wall with pins on it" (Chowka, "Original Mind," June, 37); Snyder, *Old Ways*, 39.

14. Snyder, *Earth*, 107-8, 122, 110, 112; see also Geneson, "Interview," 100-101.

15. Geneson, "Interview," 91; Whitman, *Leaves*, 230; Geneson, "Interview," 91.

16. Whitman, *Leaves*, 344.

17. Whitman uses the same image in "Thou Mother with Thy Equal Brood" (*Leaves*, 460) to describe the Civil War and America's troubles: "The livid cancer spread its hideous claws, clinging upon thy breasts, seeking to strike thee deep within, / Consumption of the worst, moral consumption, shall rouge thy face with hectic, / But thou shalt face thy fortunes, thy diseases, and surmount them all."

18. F. Scott Fitzgerald, *The Great Gatsby* (New York: Charles Scribner's Sons, 1925), 182.

19. Geneson, "Interview," 75-76. Snyder's comments on Cincinnati echo Whitman, who also found "the strange charm of aboriginal names," urged their re-adoption, and noted "among names to be revolutionized: that of the city of 'Baltimore'" (see *An American Primer*, ed. Horace Traubel [1904; reprinted San Francisco: City Lights, 1970], 30).

20. Geneson, "Interview," 72.

21. Geneson, "Interview," 93.

22. Charles Altieri, "Gary Snyder's *Turtle Island*: The Problem of Reconciling the Roles of Seer and Prophet," *Boundary* 2, no. 4 (1976): 774; Alan Williamson, "Gary Snyder: An Appreciation," *New Republic*, 173, no. 18 (1975): 28; Chowka,

"Original Mind," August, 23; Geneson, "Interview," 90; Sherman Paul, "Noble and Simple," *Parnassus* 3 (1975): 221; Geneson, "Interview," 90.

23. Snyder, "A Curse on the Men in Washington, Pentagon," in Leslie Fiedler, *The Return of the Vanishing American* (New York: Stein and Day, 1969), 87; Chowka, "Original Mind," July, 39.

24. Snyder, *Myths and Texts* (New York: Totem Press, 1960), 5.

25. Snyder, *Earth,* 92–93.

26. Frederick Jackson Turner, *The Significance of the Frontier in American History* (New York: Frederick Unger, 1963), 43; Snyder, *Songs for Gaia* (Port Townsend: Copper Canyon Press, 1979).

Penance or Perception: Spirituality and Land in the Poetry of Gary Snyder and Wendell Berry

Patrick D. Murphy

The Promised Land and the Garden represent two different ways of conceptualizing humankind's relationship to the land. The first conceives of the land as location for humans to mitigate their imperfectability and draw closer to the Godhead through doing penitential labor, earning their bread by the sweat of their brows in an environment created by their god for that very purpose. The second conceives of the land as a part of nature, which enables humans to understand their own part within that larger whole through the altering of perception to recognize the perfectability of humanity within the perfect balance of the world. The terms and the antagonists of this debate between a land of penance and a land of perfection have changed through the years, but the debate continues. In contemporary American poetry the writings of Gary Snyder and Wendell Berry, two poets whose ideas are more often compared than contrasted, exemplify this continuing debate.

Both Snyder and Berry believe that individual moral value derives from *right livelihood*, which requires a proper relationship to the land based on a clear concept of place and an abiding spirituality.[1] But they strongly differ in the presentations of that proper relationship of humanity to land, and the direction of that difference is reflected in the divergent religious traditions with which they identify and on which they build their spirituality. These divergent traditions lead Berry to the Promised Land and Snyder to the Garden.

Though both Snyder and Berry were born into the Judeo-Christian tradition of American culture, they early rejected Christian religious orthodoxy in favor of more naturalistic beliefs, which can be characterized by Snyder's conception of Gaia and Berry's conception of the Creation. Snyder presents Gaia, the pagan deity, as an embodiment of the earth as a separate being, a living, self-generating entity. Gaia comprises for Snyder a fusion of religious myth, spiritual belief, ecological science, and the Garden.[2] Berry presents the Creation, not capitalized in his earlier writings, as the product of some primary creative act by a preexistent being. A biblically derived term, Berry's Creation serves as a fusion of Christian values, freed from religious orthodoxy and church trappings, spiritual belief, ecological science, and the Promised Land. Both sets of beliefs continue to undergo change and modification, but one element clearly reveals itself: Snyder's beliefs have remained outside of and in opposition to the Judeo-Christian tradition; Berry's beliefs have gradually and more openly returned to that tradition.

Berry's movement has been toward the Christian sense of humanity's stewardship of the land. He also promotes the development of an individual's sense of rootedness in the land, a commitment to and belief in the place where a person is *fated* to live and where he cultivates his land and his neighbors.[3] Through this sense of stewardship, farming becomes a ritual activity, an accumulation of grace through agricultural and sociocultural deeds which bond him to land and community.

Berry's early poetry rarely speaks of God or religion except to denounce institutional hypocrisy, as in "Canticle" in *The Broken Ground.* But though such references appear only occasionally, they still imply a belief in the divinity of Christ, which surfaces in some of his condemnations of hypocrisy, as in "The Design of the House: Ideal and Hard Time" in *Findings* and in "The Guest" from *The Broken Ground.* In the latter poem a Christ figure is presented and the narrator thinks of himself as the Good Samaritan. Further, the capitalization of *God* in the early poems when referring to Christ and Christian themes contrasts with noncapitalization in the literary use of a nature deity, "the god of the river," suggesting some belief in the biblical deity as separate from other mythic conceptions of supernatural beings.[4]

Berry's emphasis, though, rests not on a creator but on the Creation, the earth in its natural state. Poems such as "Grace" in *Openings* present the idea of a unity between the cycle of the earth and the will of a god. To align oneself through a relationship with the land and an imitation of nature in agricultural practice is to align oneself with a spiritual goal. Berry elaborates this idea of imitation in "Window Poems" in *Openings*.

> He said: if man, who killed
> his brothers and hated himself,
> is made in the image of God,
> then surely the bloodroot,
> wild phlox, trillium and May apple
> are more truly made
> in God's image, for they have desired
> to be no more than they are,
> and they have spread each other.
> Their future
> is undiminished by their past.[5]

For Berry the way of nature is the way of God, and the practicing of agriculture on the farm can imitate nature and revere this way through nurturing and safeguarding, caring and revitalizing. Yet, Berry faces a contradiction because, although the farmer's cycles of planting and harvesting may follow the natural cycle, farming does not imitate nature. A farmer can never achieve such fecundity and diversity or afford such profligacy and indeterminacy. He must constantly compel the soil to yield to him crops of his choosing.

Berry does not resolve this contradiction but tends to circumvent it by focusing on the issue of revitalization. Humanity needs to work at revitalizing the land because, unlike the wild flowers, it has been diminished by its past. "The Morning News" and "The Wages of History" in *Farming: A Handbook* clearly enunciate this belief. Berry believes that a person can achieve grace and atone for the sins of previous generations by practicing a revitalizing agriculture. Even though he defines this atonement in naturalistic terms as "at-one-ment" with the earth, his own poems and essays develop it as an act of restitution and penance for humanity's past failures and the individual's own failings. Despite whatever institutional aspects of religion he rejects, Berry adopts from the Judeo-Christian tradition the belief that humanity lives in a postlapsarian world. He accepts at least a secular equivalent of original sin: the living must make restitution for the errors of preceding generations. America as Promised Land provides the opportunity for God's people to make such restitution primarily through agriculture. He defines this postlapsarian state and agricultural labor as penance in both naturalistic and Christian terms.

> It was his passion to be true
> to the condition of the Fall—
> to live by the sweat of his face, to eat
> his bread, assured the cost was paid.
> (*The Wheel*, 9)

Berry's more open Christian presentation of the Creation appears most frequently in his last two books of verse, *A Part* and *The Wheel*, and is reinforced by his recent prose collection, *The Gift of Good Land*. But even earlier, in the essay "Discipline and Hope," he unites the role of farmer with the Christian belief that *man* is made in the image of God—"As husbandman, a man is both the steward and the likeness of God, the greater husbandman"—and claims that this relationship stems from "an ancient system of analogies that clarifies a series of mutually defining and sustaining unities: of farmer and field, of husband and wife, of the world and God" (*Recollected Essays*, 213). The earth, then, is not sacred in its own right but only sacred insofar as it represents the Creator or contains him within its aspects, just as the land may be fertile but is not "productive" except to the degree to which the farmer works it. It should also be noted that with this Christian belief, Berry finds it impossible to avoid echoing a vestige of its Old Testament patriarchal trappings in the possessive subordination of land to *man* and woman to *man*.

Although Berry verges on acceptance of basic Christian tenets, he does not feel at ease. In "A Secular Pilgrimage," written only a few years before the publication of "Discipline and Hope," he defines his use of the word *religious* to exclude such beliefs: "And I will never use the word here to refer to any of those revealed certainties that are so large a part of the lore of the various churches."[6] He mainly objects to the otherworldliness of religions and adopts his concept of "at-one-ment" from John Collis, who uses the concept to encourage a return to a primitive "intimation of the Divine, which will link us again with *animism*" (408). One finds, however, no animism in Berry's writings, except in the literary devices of metaphor and simile; rather, one finds a vacillation between a naturalistic faith and fundamental Judeo-Christian beliefs. It seems that Berry must have been involved in an internal spiritual struggle in the early 1970s, torn between a romance nature religion and Christianity, reminiscent of the intellectual and theological debates of early nineteenth-century America.

With the publication of *A Part, The Wheel,* and *The Gift of Good Land* between 1980 and 1982, this inner conflict appears to have resolved itself in favor of Christian doctrine. Three poems stand out in *A Part:* "To the Holy Spirit," which ends "By Thy wide grace show me Thy narrow Gate"; "The Way of Pain," which speaks of Abraham's sacrifice and the Crucifixion; and "We Who Prayed and Wept," which ends, "We have failed Thy grace. / Lord, I flinch and pray, / send Thy necessity." The overt Christianity of these poems, as well as others in *The Wheel,* is reinforced by the title essay of *The Gift of Good Land,* which sets out "to attempt a Biblical argument for ecological and agricultural responsibility" (267). Here Berry attempts to repudiate Lynn White, Jr.'s essay, "The Historical Roots of Our Ecological Crisis," which argues that the biblical tradition has been one of the causes of Western civilization's ecological destruction. Berry uses the Bible not metaphorically but quite literally, accepting the idea that "the Promised Land is a divine gift to a *fallen* people" and arguing, "In the Bible's long working out of the understanding of this gift, we may find the beginning—and, by implication, the end—of the definition of an ecological discipline" (269). Berry demonstrates here his acceptance of the Bible as an inspired source for a spiritual approach to the proper relationship of humanity to the land; significantly, he designates the land as belonging to Jehovah: "God too loves material things; He invented them" (278).

Berry seeks to reform and revive the Judeo-Christian tradition through an accurate reinterpretation that would make it more *earthly,* as John Lang has correctly noted ("'Close Mystery'"), and he would demonstrate this reinterpretation through proper stewardship of the good land given to fallen man by the Creator. Berry's increasing emphasis on the "Fall," on atonement and restitution for ecological "sins," on the notion of "grace," and, finally, on the Bible as a source for spiritual guidance indicate that his definition of "religious" has changed considerably since "A Secular Pilgrimage." He has moved away from Collis's "intimation of the Divine" and ideas of animism and nature worship back toward divine revelations of the Old Testament Jehovah and the New Testament Christ, with an attendant belief in humanity's exile from the Garden and its passage into the Promised Land.

When one turns to Snyder and attempts to chart his direction in the same manner as charting Berry's, one encounters a web of paths rather than a single one departing from and returning to one tradition. Rather than toward stewardship, Snyder moves toward an immersion in natural process, a union with the land both wild and cultivated. This course includes an early repudiation of the Judeo-Christian tradition. Conveniently, Snyder also makes explicit reference to Lynn White, Jr.'s essay, but he upholds it as an example illustrating the need to overthrow the myth of the Old Testament God, "the Jehovah Imperator God-figure, a projection of the gathered power of anti-natural social forces" (*Earth House Hold,* 123–24). While still in his teens, Snyder broadened his rejection to encompass all of the mainstream Western tradition.

> When I went to college I was bedeviled already by the question of these contradictions of living in and supposedly being a member of a society that was destroying its own ground.... I began to perceive that maybe it was all of Western culture that was off the track. (*The Real Work,* 94)

American culture's relationship to the land provided Snyder not only a basis for rejecting Western culture and its religious traditions but a criterion for evaluating other traditions that might contribute to a new culture for the United States. Native American tribes provided Snyder with one such example, but he soon discovered that their traditions were not open to study: "Its content, perhaps, is universal, but you must be a Hopi to follow the Hopi way" (*The Real Work,* 17). His limited immersion in Native American studies, though, provided Snyder with a solid spiritual base for the next leg of his path: the inward journey of formal Zen training.

Two works provide a detailed presentation of the beliefs developed by Snyder during and immediately after his undergraduate studies, the years preceding his departure for Japan. *He Who Hunted Birds in His Father's Village,* written as his undergraduate honor's thesis at Reed College, presents an analysis of a Haida myth. He demonstrates in it his working out of the relationships between Native American cultural beliefs and ancient shamanism, in a sense, modern and historical primitivism. He also develops his belief in the relationship between the shaman as visionary and myth preserver in primitive cultures and the role of the modern poet as creator of new cultural myths and preserver of the ancient traditions within the present. *Myths & Texts,* a poetic sequence, presents these ideas as well as Snyder's linkages among ancient shamanism, Native American primitivism, and Hindu-Buddhist beliefs, in the form of a poetic vision quest.

In Snyder's study Native Americans form part of a nexus of myth-united cultures stretching from Samothrace to pre-Aryan India to the Americas, which have strong matrilineal and goddess-worship characteristics. He also sees vestiges of this cultural unity surfacing among the patrilineal, male-god systems which have largely replaced it. He defines these vestiges in *Earth House Hold* as The Great Subculture with a forty-thousand-year history (114–15). Snyder ascribes to the patrilineal, male-god cultures the same separation of body and mind, humans and nature, reason and spirit that Berry criticizes in his work. Essentially, Snyder synthesizes a series of anthropological, psychological, and mythological theories drawn from Jung, Campbell, Cassirer, Kerenyi, Frazer, and others into a single theory illuminating a primitive cultural-spiritual archetype strongly influenced by Robert Graves's concept of the White Goddess.[7]

On the basis of a belief in the underlying mythological unity among ancient and primitive cultural traditions, Snyder sees a relationship between the vision quests and shamanist practices of primitive cultures and Zen, indicating that as he turned toward the study of Buddhism, he had already formed a larger spiritual and philosophical framework from which to approach that study. Buddhism would serve for him as another aspect of an ongoing spritual tradition requiring new developments to enable it to move from the position of subculture back to main culture. And within this revitalization, the poet, modern descendant of the shaman, must play a vital role.

Native American tribal spiritualism and Buddhism, then, form two different spiritual paths within a single larger tradition for Snyder. The former demonstrates a direct cultural link with the "old ways," while the latter demonstrates a tortuous philosophical tradition wending its way through several national cultures and undergoing various transformations into different schools and sects. Nevertheless, Snyder believes that Zen, Tantrism, Avatamsaka, Taoism, and Vajrayana, which he claims "has traditional continuous lines that go back to the Stone Age" (*The Real Work,* 176), all have deeply archaic roots. Snyder includes gnostic and Christian sects within this tradition as well, but these, like the Native American tribes, are relatively inaccessible or extinct and incapable of acting as vehicles for a "new social mythology" in the United States (*He Who Hunted Birds*, 112; *Myths and Texts*, viii). Snyder went to Japan because he believed that Buddhism might serve as such a vehicle (*The Real Work*, 95). He views the religious disciplines of the Orient as providing an organized approach to the study of the powers achieved through meditation and yoga, powers wielded by ancient shamans but never understood or handed down by them in the manner of religious learned traditions, such as in Tantrism or Zen (*The Real Work*, 15).

Snyder always intended to return to the United States with a very specific purpose in mind: "It wasn't just returning—the next step of my own practice was to be here" (*The Real Work*, 99). This practice would be nothing short of promoting the formation of a new culture.

> What we need to do now is to take the great intellectual achievement of the Mahayana Buddhists and bring it back to a community style of life which is not necessarily monastic. Some Native American groups are a good example. (*The Real Work*, 16)

Not only did Snyder have such a goal in mind when he left for Japan, but he also had determined his own vehicle for achieving that goal: poetry. He presents this belief in *Myths & Texts:* "I sit without thoughts by the log road / Hatching a new myth" (19).[8] Significantly, he includes these lines in the section of the sequence subtitled "First Shaman Song."

It is not possible to adequately discuss *Myths & Texts* in a few pages; its complexity denies such a synopsis.[9] Julian Gitzen captures some of the interconnections woven through the book when he observes that "Buddhist compassion and Indian empathy are available as instruments for restoring Diana's rule while bringing harmony to our lives and to our natural surroundings."[10] In *Myths & Texts* Diana's rule represents the White Goddess and a reverence for all of nature. Snyder later clarifies this mythological symbol through employing Gaia, the Earth Mother, in place of Diana (*The Old Ways*, 39). But in *Myths & Texts* Diana serves an additional purpose because defense and worship of the goddess Diana signifies a direct opposition to Christianity. The epigraph from Acts refers to Christianity's threat to the worship of Diana and reverence for the sacred grove, and in the first section of the sequence, "Logging," Snyder quotes Exodus: "But ye shall destroy their altars, break their images, and cut down their groves" (3). Throughout *Myths & Texts*

Snyder attacks the rapacity of Western Judeo-Christian culture, but he also points out that this tradition of exploitation and destruction is not limited to the West, having also occurred in China and Japan (3-4). In other words, all the societies which have departed from the primitive worship of the goddess share responsibility for the cutting down of the sacred groves, the loss of a proper relationship to the land based on recognition of its inherent sacredness. *Myths & Texts* not only describes the separation of humanity from nature as a result of following false traditions but also presents the spiritual elements needed to form a new culture that can end this separation.

Myths & Texts emphasizes that this separation results from a false dichotomy of myths and texts—"symbols and sense-impressions" (vii)—which primarily requires a new vision that recognizes their genuine unity. The outward journey, contact with wilderness and land—which Berry emphasizes in his comments to Willie Reader—can provide the evidence for the reality of humanity's life on the planet, but only the inward journey of the individual can reveal that reality is composed of an identity of the physical (texts) and the spiritual (myths) in an equal, interpenetrating relationship. Claude Levi-Strauss summarizes such a relationship in *Myth and Meaning*.

> If we are led to believe that what takes place in our mind is something not substantially or fundamentally different from the basic phenomenon of life itself, and if we are led to the feeling that there is not the kind of gap which is impossible to overcome between mankind on the one hand and all other living beings—not only animals, but also plants—on the other, then perhaps we will reach more wisdom, let us say, than we think we are capable of.[11]

In *The Back Country* and in *Mountains and Rivers Without End*, particularly the section "Journeys" in the latter, Snyder further develops this vision, increasingly presenting it in terms of Zen enlightenment. Connected with this, he tends to eschew Buddhist teleology in favor of the belief of Zen mysticism that, as he says, "it's already all a Buddha right now if you can just see it.... It's the eternal moment."[12] The key to a new culture lies in the stripping away of illusions rather than in a long process of gaining merit through exemplary acts or the long process of penance for the "wages of history" in which Berry believes. This enlightenment will break down the artificial barriers that humans imagine exist between themselves and nature, and that recognition will provide the basis for an accurate concept of place—the "earth house hold"—and the working out of a proper relationship of humanity and land utilizing all the learning of the Great Subculture. The preserver and spokesperson for that subculture in the present must be the poet-shaman.

> The shaman speaks for wild animals, the spirits of plants, the spirits of mountains, of watersheds.... In the shaman's world, wilderness and the unconscious become analogous: he who knows one and is at ease in one, will be at home in the other. (*Old Ways,* 12)

Snyder emphasizes the first part of the preceding quote in *Turtle Island,* delineating this purpose in his explanation of the volume's title.

> A name: that we may see ourselves more accurately on this continent of watersheds and life communities—plant zones, physiographic provinces, culture area.... Each living being is a swirl in the flow.... The land, the planet itself, is also a living being—at another pace. ("Introductory Note")

In this collection he emphasizes ecological defense of nature and opposition to the "gathered power of anti-natural forces" identified in *Earth House Hold.* In *Axe Handles,* his most recent collection of poems, he continues to emphasize both the poet-shaman's role in handing down to the next generation the lessons of the Great Subculture that will enable it to see that "it's already all a Buddha" here in the Garden and the spiritual path of matrilineal nature-goddess worship in his continuing reverence for Gaia.

Jamake Highwater, in *The Primal Mind,* states that "it is evident to thoughtful people that Western society no longer has a viable functioning myth. It therefore has no basis to affirm life."[13] Snyder and Berry would both agree with this claim. In the work of both poets, they have sought not only to represent their concepts of place and relationships to the land but also to imbue these with the spiritual power of religious myth. The degree to which Snyder has consciously approached this task can be evidenced by his intense concern with the cultural role of myth from the time of his bachelor's thesis to the poems in *Axe Handles.* Berry's consciousness of mythopoeia is not as apparent as Snyder's, but his efforts in this direction are reflected throughout his poetry and prose. In both cases the spiritual beliefs go beyond a relationship to the land, but their key elements bear heavily on the sense of that relationship and are tested by both poets through the living of those relationships, Snyder in the Sierra Nevada foothills, and Berry on his Kentucky farm.

Berry believes that humanity lives in a postlapsarian world. "The Fall" exists as a central mythic element and outlines the character of humanity's historical fate. Berry believes that humanity is separated from nature as a result of this fall from grace and that humanity can only heal that rift through a long, slow process of restitutive labor. Only by "the sweat of his face" can a person achieve "at-one-ment" with nature again: the achievement of grace comes primarily through productive labor that respects, replenishes, and rectifies the land.

As Berry moves back into the Judeo-Christian tradition, entering into harmony with its essential conception of sin, he increasingly sees the Creation as an *act* of God rather than an entity unto itself of which humans are a part. Rather, humans are also a creation of God, who have received the land as a gift, and through properly working it they can prove themselves worthy of some grace by providing the Creator of both humans and nature "the joys of recognition" (*The Wheel,* 12). He has essentially turned in this movement toward a myth of America originated by the Puritans: the Promised Land in which a chosen people may do penance and prove themselves worthy of "the gift of good land."

Snyder does not believe in any fall from grace—"No paradise, no fall. / Only the weathering land / The wheeling sky" (*Riprap,* 8)—that requires penance or restitution. The function of labor, of experience, is to place a person back into contact with the reality of the earth as Garden, particularly in its wild, self-perpetuating state, where one can recognize the interpenetrating unity that exists among all of nature's elements, including the human. The separation of humans and nature exists only so long as people believe in it; because people do not know themselves they cannot know nature. Snyder knows, following the tradition of Emerson's "Nature," that to come to terms with either is to come to terms with both since they are the same.[14] Snyder quotes Chan-jan in *The Old Ways:* "Who then is 'animate' and who 'inanimate'? Within the Assembly of the Lotus, all are present without division" (9).

Although both Snyder and Berry promote through their writings and through the examples of their lives commitment to an integral spiritual component in modern life and to personal and social ecological responsibility in defense and promotion of the preservation of nature, they pursue differing conceptions of the right spiritual direction for American culture. They share many practical goals and a deep abiding commitment to spiritually revitalizing and reorienting American culture, as well as having mutual respect for each other, as indicated in "To Gary Snyder" in *A Part* and "Berry Territory" in *Axe Handles*. Nevertheless, their divergent positions on the spiritual tradition which should guide that cultural reorientation establish two different conceptions of America in particular, and the Earth in general, as Promised Land or as Garden. The dichotomy of these beliefs exemplifies not only the continuing debate in American culture on this point but also a crucial debate within the ranks of those who seek to preserve the earth in opposition to those who are daily destroying it.

NOTES

1. On Snyder and Berry's conceptions of a proper relationship to the land based on a clear concept of place, see my essay, "Two Different Paths in the Quest for Place: Gary Snyder, Wendell Berry," *American Poetry* 2, no. 1 (1984): 60–68.

2. In "The Politics of Ethnopoetics" (*The Old Ways*, 15–43), Snyder explicitly discusses the Gaia hypothesis, referring to Lovelock and Epton's essay, "The Quest for Gaia," *New Scientist* 65, no. 935 (1975): 304–6. Also see J. E. Lovelock, *Gaia: A New Look at Life on Earth* (Oxford: Oxford University Press, 1979) for a more detailed discussion of the Gaia hypothesis. Also see Snyder's "Introductory Note" to *Turtle Island*.

3. Bruce Williamson, "The Plowboy Interview: Wendell Berry," *Mother Earth News* 20 (1973): 9.

4. John Lang's essay somewhat misrepresents Berry's early poetry by failing to make this distinction.

5. Lang also quotes from this poem, but leaves off the last four and one-half lines, thereby deleting Berry's early acceptance of humanity's postlapsarian state, an essential aspect of his spiritual beliefs.

6. Wendell Berry, "A Secular Pilgrimage," *Hudson Review* 28 (1975): 402.

7. Specifically, see chapters 4, 5, and 6 of *He Who Hunted Birds in His Father's Village*. There he states: "I can only answer that the total view achieved by this method is my own doing, created by the selection of relevant concepts and implying no criticism of or assent to any one point of view" (98).

8. I must take issue with Lee Bartlett's reading of these lines in his essay, "Gary Snyder's *Myths & Texts* and the Monomyth," *Western American Literature* 17, no. 2 (1982): 137–48, when he states that "he sits quietly without thinking, his

unconscious 'hatching a new myth' to sustain him" (141). These last three words overemphasize the personal nature of the protagonist's quest and overlook the socially redeeming function of myth and mythopoeia as well as the sense of social responsibility implied by the piece's subtitle, "first shaman song," which links the modern poet's role to that of his primitive precursor.

9. I have treated *Myths & Texts* at length in "Alternation and Interpenetration: Gary Snyder's *Myths & Texts, Critical Essays on Gary Snyder,* ed. Patrick D. Murphy (Boston: G. K. Hall, 1991), 210–29.

10. Julian Gitzen, "Gary Snyder and the Poetry of Compassion," *Critical Quarterly* 15 (1973): 356.

11. Claude Lévi-Strauss, *Myth and Meaning* (New York: Schocken Books, 1979), 24.

12. Ekbert Faas, *Towards a New American Poetics: Essays and Interviews* (Santa Barbara, CA: Black Sparrow Press, 1971), 109.

13. Jamake Highwater, *The Primal Mind* (New York: New American Library, 1981), 41.

Afterword: Toward an Ecocriticism

John Cooley

The majority of the essays in this book engage the represented nature writers on fairly conventional grounds, with more interest in text and author than in historical or political context, and with more interest in the conventions of practical criticism than in the various poststructuralist approaches. However, important signs indicate that a developing ecocentrism may well shape critical engagement with American nature writing in the next decade. Credit for coining the term *ecocriticism* goes to William Rueckert, author of the essay on Barry Lopez in this book, for a preliminary essay on the topic published more than a decade ago.[1]

Ecofeminism and feminist readings of nature writing are already challenging the assumption of literary history and pastoralism along gender lines, as is evidenced in at least two essays included in this book. Feminist revisionism is well under way, bringing to light the work of "lost" women nature writers, and revealing that women's wilderness and frontier experiences, for example, differed sharply from those of men. Lawrence Buell aptly comments that in researching the rise of the nineteenth-century nature essay, he

> was surprised to discover the degree of interdependence between the "major" male figures and the work and commentary of women writers less well known. Roughly half the nature essays contributed to the *Atlantic Monthly* during the late nineteenth century—when the nature essay became a recognized genre—were by female authors.[2]

I have already mentioned, in the Introduction, the fresh and insightful readings of pastoral texts that can emerge when critics, such as Roger

Sales, adopt a new historicist approach, with its particular sensitivity to social and political conditions, economics, and ideology.

American nature writing is particularly ripe for engagement by the full range of contemporary theoretical approaches. Some interesting connections exist between the emerging ecocriticism and poststructuralist and postmodernist theories. (For the purpose of simple identification I shall call these approaches theory and ecology.) In general, both positions are revolutionary, questioning the dominant patriarchal, logocentric, and technocentric structures of Western culture. Both practices turn from the constituted center, with its dominant ideologies and metatheories, to the margins—from authoritative centrality to mostly powerless, but nonetheless vital and diverse, marginality. Both theory and ecology look to the vitality of "difference" rather than to the dangerous dominance of "sameness." The former often takes on the task of deconstructing metamyths—old hierarchies and canonizations. The latter values preserving species diversity and embraces circular rather than linear conceptions of thought and action. Ecology also documents nature's inevitable preference for diversity rather than monopoly.

Ecocritics will find some fertile common ground with contemporary literary theorists. Both camps embrace an organic or ecological conception of textuality, viewing any given text as a complex community of intrinsic and extrinsic relationships. Once a (nature writing) text is seen as an organic community of interrelated entities, ranging from conditions of production to conditions of reception, criticism shifts its emphasis toward a systemic study of textual interrelationships instead of reading primarily to produce meaning.

The practices of contemporary theory should encourage ecocriticism to question its assumptions about objectivity. Political criticism offers many useful reminders that authority always poses as objective. Ecocritics will need to examine the subjective roots of even our seemingly most scientific and objective "truths." As Gary Zukhav observes in *The Dancing Wu Li Masters:* "There is no such thing as objectivity. We cannot eliminate ourselves from the picture. We are part of nature, and when we study nature there is no way around the fact that nature is studying itself."[3] All texts, including critical essays, are subjective and contextually situated with regard to many variables, including culture, politics, economics, and history. Theory reminds us that despite our most objective intentions, we are bound in our readings and writings by our situatedness. As we read, we create meaning and bring the text to a new life, even if to a situated and subjective one.

Theory and ecology will also find much in common, I suspect, in the areas of intertextuality and biocentrism. Critics Harold Bloom and Julia Kristeva, among others, have explored the relationship of text to text, the web of textual cross-references and influences that form a matrix of intertextuality. The parallels between intertextuality and ecological concepts of biotic community interactions are numerous. As Fritjof Capra puts it, the world is "a complex web of relationships between the various parts and the whole."[4] In *A Sand County Almanac,* Aldo Leopold enunciates his "land ethic" that has been almost universally accepted as a basis for sound ecological thinking and practices. His ethic is already a lynchpin in ecocritical thinking about texts, and it could give wise counsel for reinforcement to critical theory. Leopold asserts that our actions are best directed toward the preservation of the "integrity, stability and beauty of the biotic community" of which we are a vital part.[5] A healthy textual environment, much like a healthy biosphere, is best promoted by encouraging conditions favorable for the production and reception of a numerous and diverse textuality. In both ecological and (poststructuralist) textual practices, a healthy environment preserves maximum freeplay of related elements, is intensely aware of the situatedness of a text or biotic community, and resists extensive utilization for specialized purposes, or in poststructuralist terms, "resists closure."

Neither texts nor biotic communities are closed systems. Thus, theory and ecology can both be expected to value textual and biological diversity. The health of the biosphere, and presumably of literary textuality, is best served by practices that enable distinct biomes and constituent communities to continue their evolutionary activity by various forms of interaction and change, such as trial and error, adaptation, predation, and mutualism. On this latter point, ecology may challenge theory to recognize the primary role of biological systems, against which all anthropocentric activities and values, including the production of language and texts, play a minor and dependent role.[6]

NOTES

1. See William Rueckert, "Literature and Ecology: Experiment in Ecocriticism," *Iowa Review* 9, no. 1 (Winter 1978): 71–86.

2. See Lawrence Buell, "American Pastoral Ideology Reappraised," *Norton Anthology of American Literature* (New York: W. W. Norton: 1992), 463–79.

3. Gary Zukav, *The Dancing Wu Li Masters: An Overview of New Physics* (New York: Bantam, 1980), 31.

4. Fritjof Capra, *The Tao of Physics* (Oxford: Fontana, 1976), 71.

5. Aldo Leopold, *A Sand County Almanac, With Essays on Conservation from Round River* (New York: Ballantine, 1970), 262.

6. Sue Ellen Campbell, "The Land of Language and Desire: Where Deep Ecology and Post-Structuralism Meet," *Western American Literature* 24, no. 3 (1989): 199–211 (a most helpful and influential essay).

Contributors

Edward Abbey (1927–89) published thirty books and numerous articles during his long and notable literary career. His writing betrays his powerful attraction to wilderness and distrust of American society.

Thomas C. Bailey is associate professor of English at Western Michigan University. In addition to his interest in John McPhee, he is currently writing on Edward Hoagland and Annie Dillard.

Wendell Berry writes poetry, fiction, and essays and lives and farms in Port Royal, Kentucky. His work reveals a love of nature and agriculture, a strong sense of place and community.

John Cooley is professor of English and Environmental Studies at Western Michigan University. His previous writing includes *The Great Unknown: The Journals of the First Expedition of the Colorado River* and *Mark Twain's Aquarium: The Samuel Clemens–Angelfish Correspondence.*

Ed Folsom is professor and chair of the English Department at the University of Iowa. He is editor of the *Walt Whitman Quarterly Review* and of books on Whitman and W. S. Merwin and author of the forthcoming book *Native Representations in Whitman's Leaves of Grass.*

Jack Hicks is professor of English at the University of California at Davis.

James I. McClintock is professor of American Studies at Michigan State University. His latest book, *Nature's Kindred Spirits,* is forthcoming from the University of Wisconsin Press.

Gary McIlroy teaches at Henry Ford Community College in Dearborn, Michigan. He has written articles on a variety of natural history subjects.

Patrick D. Murphy is professor and chair of the English department at Indiana University of Pennsylvania. He is founding editor of the journal *Isle: Interdisciplinary Studies of Literature and Environment* and has published two books on Gary Snyder.

William H. Rueckert is professor emeritus of English at the State University of New York at Geneseo. He has written extensively on topics relating to literature and ecology and has published books on Glenway Wescot and Kenneth Burke.

John Tallmadge writes extensively about nature writers and teaches at the Union Institute in Cincinnati.

Diane Wakoski teaches at Michigan State University and is well known for her distinctive poetic voice and vision. She is frequently named among major contemporary poets, and she has won important literary prizes, including a Robert Frost Fellowship and a Guggenheim Grant.

Steven Weiland is professor of Higher Education at Michigan State University, where he teaches courses on adult development and aging and on the history and structure of academic disciplines.

Bibliographies

Edward Abbey: Works

Abbey's Road. New York: E. P. Dutton, 1979.
The Best of Edward Abbey. San Francisco: Sierra Club Books, 1984.
Beyond the Wall: Essays from the Outside. New York: Holt, Reinhart and Winston, 1984.
Black Sun. Santa Barbara, CA: Capra Press, 1981.
The Brave Cowboy: An Old Tale in a New Time. New York: Dodd, Mead, and Co., 1956.
Cactus Country. New York: Time-Life Books, 1973.
Confessions of a Barbarian. Santa Barbara, CA: Capra Press, 1986.
Desert Images: An American Landscape. With David Muench. New York: Harcourt, Brace Jovanovich, 1979.
Desert Solitaire: A Season in the Wilderness. New York: McGraw-Hill, 1968.
Down the River. New York: E. P. Dutton, 1982.
Fire on the Mountain. Albuquerque: University of New Mexico Press, 1962.
The Fool's Progress: An Honest Novel. New York: Holt, Reinhart and Winston, 1988.
Good News. New York: E. P. Dutton, 1980.
Hayduke Lives: A Novel. Boston: Little, Brown, and Co., 1990.
The Hidden Canyon: A River Journey. With John Blaustein. New York: Penguin Books, 1977.
Jonathan Troy. New York: Dodd, Mead, and Co., 1954.
The Journey Home: Some Words in Defense of the American West. New York: E. P. Dutton, 1977.
The Monkey Wrench Gang. Philadelphia: J. B. Lipincott, 1975.
One Life at a Time, Please. New York: Henry Holt and Company, 1978.
Slickrock: The Canyon Country of Southeast Utah. With Philip Hyde. San Francisco: Sierra Club Books, 1971.
Slumgullion Stew: An Edward Abbey Reader. New York: E. P. Dutton, 1984.
Sunset Canyon: A Novel. London: Talmay, Franklin, 1971.

A Voice Crying in the Wilderness (Vox Clamantis in Deserto) Notes From a Secret Journal. New York: St. Martin's Press, 1989.

Edward Abbey: Bibliography

Alton, James Martin. "'Sons and Daughters of Thoreau': The Spiritual Quest in Three Contemporary Nature Writers." *DAI* 42 (1981): 701A.

Benoit, Raymond. "Again with Fair Creation: Holy Places in American Literature." *Prospects: An Annual Journal of American Cultural Studies* 5 (1980): 315–30.

Greiner, Patricia. "Radical Environmentalism in Recent Literature Concerning the American West." *Rendezvous: Journal of Arts and Letters* 19, no. 1 (1983): 8–15.

Hepworth, James, and Gregory McNamee, eds. *Resist Much, Obey Little: Some Notes on Edward Abbey.* Salt Lake City, Utah: Dream Garden Press, 1985.

Herndon, Jerry A. "'Moderate Extremism': Edward Abbey and 'The Moon-eyed Horse.'" *Western American Literature* 16, no. 2 (1981): 97–103.

McCann, Garth. *Edward Abbey.* Boise, Idaho: Boise State University Press, 1977.

McClintock, James. "Edward Abbey's 'Antidotes to Despair.'" *Critique,* Fall 1989, 41–54.

Peterson, Levi S. "The Primitive and the Civilized in Western Fiction." *Western American Literature* 1, no. 3 (1966): 197–207.

Pilkington, William T. "Western Philosopher, or How to be a 'Happy Hopi Hippie.'" *Western American Literature* 9, no. 1 (1974): 17–31.

Ronald, Ann. *The New West of Edward Abbey.* Albuquerque: University of New Mexico Press, 1982.

Standiford, Les. "Desert Places: An Exchange With Edward Abbey." *Western Humanities Review* 24, no. 2 (1970): 395–98.

Twining, Edward S. "Edward Abbey, American: Another Radical Conservative." *Denver Quarterly: A Journal of Modern Culture* 12, no. 4 (1978): 3–15.

Wylder, Delbert E. "Edward Abbey and the 'Power Elite.'" *Western Review* 6 (Winter 1969): 18–22.

Wendell Berry: Works

The Broken Ground. New York: Harcourt Brace Jovanovich, 1964.

Clearing. New York: Harcourt Brace Jovanovich, 1977.

Collected Poems, 1957–1982. San Francisco: North Point Press, 1985.

A Continuous Harmony: Essays Cultural and Agricultural. New York: Harcourt Brace Jovanovich, 1972.

The Country of Marriage. New York: Harcourt Brace Jovanovich, 1973.

An Eastward Look. Berkeley, CA: Sand Dollar, 1974.

Farming: A Hand Book. New York: Harcourt Brace Jovanovich, 1970.

Fidelity: Five Stories. New York: Pantheon Books, 1992.
Findings. Iowa City, IA: Prairie Press, 1969.
The Gift of Good Land: Further Essays, Cultural and Agricultural. San Francisco: North Point Press, 1981.
The Hidden Wound. Boston: Houghton Mifflin, 1970.
Home Economics. San Francisco: North Point Press, 1987.
Horses. Monterey, KY: Larkspur Press, 1975.
Harlan Hubbard: Life and Work. Lexington, KY: University Press of Kentucky, 1990.
The Kentucky River: Two Poems. Monterey, KY: Larkspur Press, 1976.
The Long-Legged Horse. New York: Harcourt Brace Jovanovich, 1969.
Meeting the Expectations of the Land: Essays in Sustainable Agriculture and Stewardship. San Francisco: North Point Press, 1984.
The Memory of Old Jack. New York: Harcourt Brace Jovanovich, 1974.
Nathan Coulter: A Novel. Boston: Houghton Mifflin, 1960.
November Twenty Six, Nineteen Hundred Sixty Three. New York: Braziller, 1964.
Openings. New York: Harcourt Brace Jovanovich, 1968.
A Part. San Francisco: North Point Press, 1983.
A Place on Earth: A Novel. New York: Harcourt Brace Jovanovich, 1967.
Recollected Essays, 1965-1980. San Francisco: North Point Press, 1981.
Remembering: A Novel. San Francisco: North Point Press, 1988.
The Rise. Lexington, KY: Graves Press, 1968.
Sayings and Doings. Lexington, KY: Gnomon, 1975.
Standing By Words: Essays. San Francisco: North Point Press, 1983.
There Is Singing Around Me. Cold Mountain Press, 1976.
Three Memorial Poems. Berkeley, CA: Sand Dollar, 1976.
To What Listens. Crete, NE: Best Cellar, 1975.
Travelling at Home. The Press of Appletree Alley, 1988.
The Unforseen Wilderness: An Essay on Kentucky's Red River Gorge. Lexington, KY: University Press of Kentucky, 1971.
The Unsettling of America: Culture and Agriculture: San Francisco: Sierra Club Books, 1977.
What Are People For? San Francisco: North Point Press, 1990.
The Wheel: Poems. San Francisco: North Point Press, 1982.
The Wild Birds: Six Stories of the Port William Membership. San Francisco: North Point Press, 1986.

Wendell Berry: Bibliography

Collins, Robert Joseph. "A Secular Pilgrimage: Nature, Place and Morality in the Poetry of Wendell Berry." *DAI* 39 (1979): 4935A.
Cornell, Daniel T. "Practicing Resurrection: Wendell Berry's Georgic Poetry, an Ecological Critique of American Culture." *DAI* 47 (1986): 951A.

Dietrich, Mary. "Our Commitment to the Land." *Bluegrass Literary Review* 2, no. 1 (1980): 39–44.
Ditsky, John M. "Wendell Berry: Homage to the Apple Tree." *Modern Poetry Studies* 2 (1971): 7–15.
Ehrlich, Arnold W. "PW Interviews: Wendell Berry." *Publishers Weekly*, 5 September 1977, 10–11.
Fields, Kenneth. "The Hunter's Trail: Poems by Wendell Berry." *Iowa Review* 1 (Winter 1970): 90–99.
Hicks, Jack. "A Wendell Berry Checklist." *Bulletin of Bibliography and Magazine Notes* 37, no. 3 (1980): 127–31.
Lang, John. "'Close Mystery': Wendell Berry's Poetry of Incarnation." *Renascence: Essays on Value in Literature* 35, no. 4 (1983): 258–68.
Logsdon, Gene. "Back to the Land." *Farm Journal*, March 1972, 30–32.
Morgan, Speer. "Wendell Berry: A Fatal Singing." *Southern Review* 20 (1974): 865–77.
Murphy, Patrick D. "Penance or Perception: Spirituality and Land in the Poetry of Gary Snyder and Wendell Berry." *Sagetrieb: A Journal Devoted to Poets in the Pound-H*, Fall 1986, 61–72.
Rodale, Robert. "The Landscape of Poetry." *Organic Farming and Gardening*, April 1976, 46–52.
Swann, B. "The Restoration of Vision." *Commonweal*, 6 June 1986, 345–46.
Tolliver, Gary Wayne. "Beyond Pastoralism: Wendell Berry and a Literature of Commitment." *DAI* 39 (1979): 6767A–68A.
Waage, Frederick O. "Wendell Berry's History." *Contemporary Poetry: A Journal of Criticism* 3, no. 3 (1978): 21–46.

Annie Dillard: Works

An American Childhood. New York: Harper and Row, 1987.
Ed. *The Best American Essays*. New York: Ticknor and Fields, 1988.
Encounters with Chinese Writers. Middletown, CT: Wesleyan University Press, 1984.
Holy the Firm. New York: Harper and Row, 1977.
The Living. New York: Harper Collins Publishers, 1992.
Living by Fiction. New York: Harpers Magazine Press, 1982.
Pilgrim at Tinker Creek. New York: Harper and Row, 1974.
Teaching a Stone to Talk: Expedition and Encounters. New York: Harper and Row, 1982.
Three By Annie Dillard. New York: Harper and Row, 1990.
Tickets for a Prayer Wheel. New York: Harper and Row, 1974.
The Writing Life. New York: Harper and Row, 1989.

Annie Dillard: Bibliography

Alton, James Martin. "'Sons and Daughters of Thoreau': The Spiritual Quest in Three Contemporary Nature Writers." *DAI* 42 (1981): 701A.

Dunn, Robert. "The Artist as Nun: Theme, Tone and Vision in the Writings of Annie Dillard," *Studia Mystica* 1, no. 4 (1978): 18.

Hammond, Karla M. "Drawing the Curtains: An Interview with Annie Dillard." *Bennington Review* 10 (April 1981): 30–38.

Lavery, David L. "Noticer: The Visionary Art of Annie Dillard." *Massachusetts Review* 21, no. 2 (1980): 255–70.

McConahay, Mary Davidson. "'Into the Bladelike Arms of God': The Quest for Meaning Through Symbolic Language in Thoreau and Annie Dillard." *Denver Quarterly* 20, no. 2 (1985): 103–16.

McFadden-Gerber, Margaret. "The I in Nature." *American Notes and Queries* 16, no. 1 (1977): 3–5.

McIlroy, Gary. "*Pilgrim at Tinker Creek* and the Burden of Science." *American Literature: A Journal of Literary History, Criticism, and Bibliography* 59, no. 1 (1987): 71–84.

———. "'The Sparer Climate for Which I Longed': *Pilgrim at Tinker Creek* and the Spiritual Imperatives of Fall." *Thoreau Quarterly: A Journal of Literary and Philosophical Studies* 16, nos. 3–4 (1984): 156–61.

———. "Transcendental Prey: Stalking the Loon and the Coot." *Thoreau Society Bulletin* 176 (Summer 1986): 5–6.

Peterson, Eugene H. "Annie Dillard: With Her Eyes Open." *Theology Today* 43, no. 2 (1986): 178–91.

Reimer, Margaret Loewen. "The Dialectical Vision of Annie Dillard's *Pilgrim at Tinker Creek.*" *Critique: Studies in Modern Fiction* 24, no. 3 (1983): 182–91.

Scheick, William J. "Annie Dillard: Narrative Fringe." In *Contemporary American Women Writers: Narrative Strategies,* ed. Catherine Rainwater and William J. Scheick, 51–67. Lexington: University Press of Kentucky, 1985.

Waage, Frederick G. "Alther and Dillard: The Appalachian Universe." In *Appalachia/America: Proceedings of the 1980 Appalachian Studies Conference,* ed. Wilson Somerville, 200–208. Johnson City, TN: Appalachian Consortium, 1981.

Yancey, Philip. "A Face Aflame: An Interview with Annie Dillard." *Christianity Today,* 5 May 1978, 14–19.

Joseph Wood Krutch: Works

The American Drama Since 1918: An Informal History. New York: Random House, 1939.

And Even if You Do: Essays on Man, Manners and Machines. New York: Morrow, 1967.
Baja California and the Geography of Hope. San Francisco: Sierra Club Books, 1967.
The Best Nature Writing of Joseph Wood Krutch. New York: Morrow, 1969.
The Best of Two Worlds. New York: Sloane, 1953.
Biology and Humanism. San Francisco: Industrial Indemnity, 1960.
The Comedies of William Congreve. New York: Macmillan, 1927.
Comedy and Conscience After the Restoration. New York: Columbia University Press, 1924.
The Desert Year. New York: Sloane, 1952.
Edgar Allen Poe: A Study in Genius. London: Alfred A. Knopf, 1926.
Experience and Art: Some Aspects of the Esthetics of Literature. New York: Harrison Smith and R. Haas, 1932.
Five Masters: A Study in the Mutations of the Novel. New York: J. Cape and H. Smith, 1930.
The Forgotten Peninsula: A Naturalist in Baja California. New York: Sloane, 1961.
Grand Canyon: Today and All Its Yesterdays. New York: W. Sloane Associates, 1961.
Ed. *Great American Nature Writing*. New York: W. Sloane Associates, 1950.
The Great Chain of Life. Boston: Houghton Mifflin, 1956.
Henry David Thoreau. New York: W. Sloane Associates, 1948.
Herbal. New York: G.P. Putnam's Sons, 1965.
Human Nature and the Human Condition. New York: Random House, 1959.
If You Don't Mind My Saying So . . . Essays on Man and Nature. New York: W. Sloane Associates, 1964.
Is the Common Man too Common. Norman: University of Oklahoma Press, 1954.
A Krutch Omnibus: Forty Years of Social and Literary Criticism. New York: Morrow, 1970.
The Last Boswell Paper. Woodstock, VT: Elm Tree, 1951.
The Measure of Man: On Freedom, Human Values, Survival, and the Modern Temper. Indianapolis, IN: Bobbs-Merrill, 1954.
Merely a Humanist. San Francisco: Industrial Indemnity, 1968.
The Modern Temper: A Study and a Confession. New York: Harcourt Brace Jovanovich, 1929.
"Modernism" in Modern Drama: A Definition and an Estimate. Ithaca, NY: Cornell University Press, 1953.
More Lives Than One. New York: W. Sloane Associates, 1962.
The Most Wonderful Animals That Never Were. Boston: Houghton Mifflin, 1969.
Samual Johnson. New York: W. Holt and Company, 1944.
Ed. *Thoreau: Walden and Other Writings*. New York: Bantam Books, 1962.
A Treasury of Birdlore. Garden City, NY: Doubleday and Co., 1962.
The Twelve Seasons: A Perpetual Calendar for the Country. New York: W. Sloane Associates, 1955.

The World of Animals: A Treasury of Lore, Legend, and Literature by Great Writers and Naturalists From 5th Century B.C. to the Present. New York: Simon and Schuster, 1961.

Joseph Wood Krutch: Bibliography

Gorman, John. "Joseph Wood Krutch: A Cactus Walden." *Melus: The Journal for the Study of the Multi-Ethnic Literature of the United States* 11, no. 4 (1984): 93–101.
Lehman, Anthony L. "Joseph Wood Krutch: A Selected Annotated Bibliography of Primary Sources." *Bulletin of Bibliography and Magazine Notes* 41, no. 2 (1984): 74–80.
Margolis, John D. *Joseph Wood Krutch: A Writer's Life*. Knoxville: University of Tennessee Press, 1980.
———. "Joseph Wood Krutch: A Writer's Passage Beyond the Modern Temper." In *Romantic and Modern: Revaluations of Literary Tradition*, ed. G. Bornstein, 135–53. Pittsburgh: University of Pittsburgh Press, 1977.
Pavich, Paul N. *Joseph Wood Krutch*. Boise, Idaho: Boise State University, 1989.
———. "Joseph Wood Krutch: Persistent Champion of Man and Nature." *Western American Literature* 13, no. 2 (1978): 151–58.
Powell, Lawrence Clark. "*The Desert Year:* Joseph Wood Krutch." In *Southwest Classics: The Creative Literature of the Arid Lands: Essays on the Books and Their Writers*, 38–53. Los Angeles: Ward Ritchie Press, 1974.
———. "Dr. Krutch's Southwest." In *The Little Package: Pages on Literature and Landscape from a Traveling Bookman's Life*. Cleveland: World Publishing Company, 1964.
Slater, Peter Gregg. "The Negative Secularism of *The Modern Temper:* Joseph Wood Krutch." *American Quarterly* 33, no. 2 (1981): 185–205.
Wild, Peter. "Ash Heap for Bright Ruderals: Joseph Wood Krutch's *The Modern Temper*." *North Dakota Quarterly* 57, no. 4 (1989): 14–22.

Aldo Leopold: Works

Aldo Leopold's Wilderness: Selected Early Writings by the Author of a Sand County Almanac. Ed. David E. Brown and Neil B. Carmony. Harrisburg, Pa: Stackpole Books, 1990.
Game Management. New York: Charles Scribner's Sons, 1933.
Report on a Game Survey of the North Central States. Madison: Democrat Printing Company, 1931.
The River of the Mother of God and Other Essays. Ed. Susan L. Flader and J. Baird Callicott. Madison: University of Wisconsin Press, 1991.

Round River: From the Journals of Aldo Leopold. Ed. Luna B. Leopold. New York: Oxford University Press, 1953.

A Sand County Almanac, and Sketches Here and There. New York: Oxford University Press, 1949.

A Sand County Almanac with Essays on Conservation from Round River. New York: Ballantine Books, 1968.

Sawdust. Frankfort, KY: Whippoorwill, 1983.

Et al. *Symposium on Hydrobiology.* Madison: University of Wisconsin Press, 1941.

Aldo Leopold: Bibliography

Aldo Leopold: The Man and His Legacy, n.a. Ankeney, Iowa: The Soil Conservation Society of America, 1987.

Attfield, Robin. *The Ethics of Environmental Concern.* New York: Columbia University Press, 1983: 158–60.

Callicott, J. Baird. "Animal Liberation: A Triangular Affair." *Environmental Ethics* 2 (Winter 1980): 311–38.

———., ed. *Companion to Sand County Almanac: Interpretive and Critical Essays.* Madison: University of Wisconsin Press, 1987.

———. "Hume's Is/Ought Dichotomy and the Relation of Ecology to Leopold's Land Ethic." *Environmental Ethics* 4 (Summer 1982): 163–74.

Flander, Susan. *Thinking Like a Mountain: Aldo Leopold and the Evolution of an Ecological Attitude Toward Deer, Wolves, and Forests.* Columbia: University of Missouri Press, 1974.

Haney, Deanna Kay. "The Role of the Literary Naturalist in American Culture." *DAI* 38 (1977): 1387A–1388A.

Heffernan, James D. "The Land Ethic: A Critical Appraisal." *Environmental Ethics* 4 (Fall 1982): 235–47.

Lehmann, Scott. "Do Wildernesses Have Rights." *Environmental Ethics* 3, no. 2 (1981): 129–47.

Moline, Jon N. "Aldo Leopold and the Moral Community." *Environmental Ethics* 8 (Summer 1986): 99–120.

Partridge, Ernest. "Are We Ready for an Ecological Morality?" *Environmental Ethics* 4 (Summer 1982): 175–190.

Williams, Bernard. "A Critique of Utilitarianism." In *Utilitarianism: For and Against,* ed. J. J. C. Smart and Bernard A. C. Williams, 77–155. Cambridge: Cambridge University Press, 1973.

Barry Lopez: Works

Arctic Dreams: Imagination and Desire in a Northern Landscape. New York: Charles Scribner's Sons, 1986.

Crossing Open Ground. New York: Charles Scribner's Sons, 1988.
Crow and Weasel. Illustrated by Tom Pohrt. San Francisco: North Point Press, 1990.
Desert Notes: Reflections in the Eye of the Raven. Mission, KS: Andrews and McMeel, 1976.
Giving Birth to Thunder, Sleeping with His Daughter. Mission, KS: Andrews and McMeel, 1977.
Of Wolves and Men. New York: Charles Scribner's Sons, 1978.
The Rediscovery of North America. Lexington: University of Kentucky Press, 1990.
River Notes: The Dance of Herons. Mission, KS: Andrews and McMeel, 1979.
Winter Count. Mission, KS: Andrews and McMeel, 1981.

Barry Lopez: Bibliography

Alton, Jim. "An Interview with Barry Lopez." *Western American Literature* 21, no. 1 (1986): 3–17.
Bonetti, Kay. "An Interview with Barry Lopez." *Missouri Review* 11, no. 3 (1988): 57–77.
Gonzalez, Ray. "The Literature of Hope: An Interview with Barry Lopez." *Bloomsbury Review* 10, no. 3 (1990): 8–9, 29.
Leuders, Edward, ed. "Dialogue One: Ecology and the Human Imagination, Barry Lopez and W. O. Wilson, Feburary 1, 1988." In *Writing Natural History: Dialogues with Authors,* 7–35. Salt Lake City: University of Utah Press, 1989.
Miller, David. "A Sense of Harmony." *Progressive* 44 (April 1980): 58–59.
Margolis, Kenneth. "Paying Attention: An Interview with Barry Lopez." *Orion Nature Quarterly* 9, no. 3 (Summer 1990): 50–53.
Nunnally, Patrick. "An Interview with Barry Lopez." *North Dakota Quarterly* 56, no. 1 (Winter 1988): 98–107.
O'Connell, Nicholas. "Interview with Northwest Writers: Barry Lopez." *Seattle Review* 8, no. 2 (Fall 1985): 29–38.
Paul, Sherman. "Making the Turn: Rereading Barry Lopez." In *Hewing to Experience,* 345–85. Iowa City: University of Iowa Press, 1989. Reprinted in *For Love of the World: Essays on Nature Writers,* 67–107. Iowa City: University of Iowa Press, 1992.
Wild, Peter. *Barry Lopez.* Boise, Idaho: Boise State University, 1984.

Peter Matthiessen: Works

African Silences. New York: Random House, 1991.
At Play in the Fields of the Lord. New York: Random House, 1965.
Blue Meridian: The Search for the Great White Shark. New York: Random House, 1971.

The Cloud Forest: A Chronicle of the South American Wilderness. New York: Viking Press, 1961.
Far Tortuga. New York: Random House, 1975.
In the Spirit of Crazy Horse. New York: Viking Press, 1983.
Indian Country. New York: Viking Press, 1984.
Killing Mister Watson. New York: Random House, 1990.
Men's Lives. New York: Random House, 1986.
Nine-Headed Dragon River: Zen Journals 1969–1985. Boston: Shambhala, 1985.
On the River Styx and Other Stories. New York: Random House, 1989.
Oomingmak: The Expedition to the Musk Ox Island in the Bering Sea. New York: Hastings House, 1967.
Partisans: A Novel. New York: Viking Press, 1955.
Race Rock: A Novel. New York: Harper and Row, 1954.
Raditzer. New York: Viking Press, 1961.
Sal Si Puedes: Cesar Chavez and the New American Revolution. New York: Random House, 1969.
Sand Rivers. New York: Viking Press, 1981.
Seal Pool. Garden City, N.Y.: Doubleday and Co., 1972.
The Shorebirds of North America. New York: Viking Press, 1967.
The Snow Leopard. New York: Viking Press, 1978.
The Tree Where Man Was Born. New York: E. P. Dutton, 1972.
Under the Mountain Wall: A Chronicle of Two Seasons in the Stone Age. New York: Viking Press, 1962.
Wildlife in America. New York: Viking Press, 1959.
The Wind Birds. New York: Viking Press, 1973.

Peter Matthiessen: Bibliography

Alton, James Martin. "'Sons and Daughters of Thoreau': The Spiritual Quest in Three Contemporary Nature Writers." *DAI* 42 (1981): 701A.
Bender, Bert. "*Far Tortuga* and American Sea Fiction Since *Moby-Dick.*" *American Literature* 56, no. 2 (1984): 227–48.
———. *Sea Brothers: The Tradition of American Sea Fiction from Moby-Dick to the Present,* chapters 11–12. Philadelphia: University of Pennsylvania Press, 1988.
Cooley, John R. "Waves of Change: Peter Matthiessen's Caribbean." *Environmental Review,* Fall 1987, 223–30.
Dowie, William. *Peter Matthiessen.* Boston: (Twayne) E. K. Hall, 1991.
Grove, James P. "Pastoralism and Anti-Pastoralism in Peter Matthiessen's *Far Tortuga.*" *Critique: Studies in Modern Fiction* 21, no. 2 (1979): 15–29.
———. "A Peter Matthiessen Checklist." *Critique: Studies in Modern Fiction* 21, no. 2 (1979): 30–38.

Heim, Michael. "The Mystic and the Myth: Thoughts on *The Snow Leopard.*" *Studia Mystica* 4, no. 2 (1981): 3-9.
Nicholas, D. *Peter Matthiessen: A Bibliography, 1951-1979.* Candea Park, CA: Orirana Press, 1979.
Patterson, Richard F. "*At Play in the Fields of the Lord:* The Imperialist Idea and the Discovery of Self." *Critique: Studies in Modern Fiction* 21, no. 2 (1979): 5-14.
———. "Holistic Vision and Fictional Form in Peter Matthiessen's *Far Tortuga.*" *Rocky Mountain Review* 37, nos. 1-2 (1983): 70-81.
Plimpton, George. "The Craft of Fiction in *Far Tortuga:* Peter Matthiessen." *Paris Review* 60 (1974): 79-82.
Styron, William. "Peter Matthiessen." In *This Quiet Dust and Other Writings,* 2-27. New York: Random House, 1982.

John McPhee: Works

Basin and Range. New York: Farrar, Straus, and Giroux, 1980.
Coming into the Country. New York: Farrar, Straus and Giroux, 1977.
The Control of Nature. New York: Farrar, Straus and Giroux, 1989.
The Crofter and the Laird. New York: Farrar, Straus and Giroux, 1970.
The Curve of Binding Energy. New York: Farrar, Straus and Giroux, 1974.
The Deltoid Pumpkinseed. New York: Farrar, Straus and Giroux, 1973.
Encounters with the Archdruid. New York: Farrar, Straus and Giroux, 1971.
Giving Good Weight. New York: Farrar, Straus and Giroux, 1979.
The Headmaster: Frank L. Boyden of Deerfield. New York: Farrar, Straus and Giroux, 1966.
Heirs of General Practice. New York: Farrar, Straus and Giroux, 1984.
In Suspect Terrain. New York: Farrar, Straus and Giroux, 1983.
The John McPhee Reader. Ed. William L. Howarth. New York: Farrar, Straus and Giroux, 1976.
La Place de la Concorde Suisse. New York: Farrar, Straus and Giroux, 1984.
Levels of the Game. New York: Farrar, Straus and Giroux, 1969.
Oranges. New York: Farrar, Straus and Giroux, 1967.
Pieces of the Frame. New York: Farrar, Straus and Giroux, 1975.
The Pine Barrens. New York: Farrar, Straus and Giroux, 1968.
Rising from the Plains. New York: Farrar, Straus and Giroux, 1986.
A Roomful of Hovings and Other Profiles. New York: Farrar, Straus and Giroux, 1968.
A Sense of Where You Are: A Profile of William Warren Bradley. New York: Farrar, Straus and Giroux, 1978.
The Survival of the Bark Canoe. New York: Farrar, Straus and Giroux, 1975.
Table of Contents. New York: Farrar, Straus and Giroux, 1985.

John McPhee: Bibliography

Clark, Joanne K. "The Writings of John Angus McPhee: A Selected Bibliography." *Bulletin of Bibliography and Magazine Notes* 38, no. 1 (January–March 1981): 45–51.
Hamilton, Joan. "An Encounter with John McPhee." *Sierra*, May/June 1990, 50–55, 92, 96.
Lounsberry, Barbara. *The Art of Fact: Contemporary Artists of Nonfiction*. Westbury, CT: Greenwood Press, 1990.
Roundy, Jack. "Crafting Fact: Formal Devices in the Prose of John McPhee." In *Literary Nonfictions: Theory, Criticism, Pedagogy*, ed. Chris Anderson, 87–99. Carbondale, IL: Southern Illinois University Press, 1989.

Gary Snyder: Works

Axe Handles: Poems. Berkeley, CA: North Point Press, 1983.
The Back Country. New York: New Directions, 1968.
Earth House Hold: Technical Notes & Queries to Fellow Dharma Revolutionaries. New York: New Directions, 1969.
The Fudo Trilogy. Berkeley, CA: Shaman Drum, 1973.
He Who Hunted Birds in His Father's Village: The Dimensions of a Haida Myth. Bolinas, CA: Grey Fox Press, 1979.
Left Out in the Rain: New Poems, 1947–1985. Berkeley, CA: North Point Press, 1986.
Myths & Texts. New York: Totem Press, 1960; 1975.
No Nature: New & Selected Poems. New York: Pantheon Books, 1992.
The Old Ways: Six Essays. San Francisco: City Lights Books, 1977.
On Bread & Poetry: A Panel Discussion between Gary Snyder, Lew Welch & Philip Whalen. Bolinas, CA: Grey Fox Press, 1977.
Passage through India. San Francisco: Grey Fox Press, 1984.
The Practice of the Wild: Essays. Berkeley, CA: North Point Press, 1990.
A Range of Poems. London: Fulcrum Press, 1966.
The Real Work: Interviews and Talks 1964–1979. New York: New Directions, 1980.
Regarding Wave. New York: New Directions, 1970.
Riprap, & Cold Mountain Poems. San Francisco: Four Seasons Foundation, 1965.
Six Sections From Mountains and Rivers Without End. San Francisco: Four Seasons Foundation, 1965.
Sons of Gaia. Port Townsend, NY: Copper Canyon Press, 1979.
Turtle Island. New York: New Directions, 1974.
War Poems. New York: The Poet's Press, 1968.

Gary Snyder: Bibliography

Allen, Frank. "East and West: Works of Robert Lowell and Gary Snyder." *Chandrabhaga* 1 (1979): 34–56.

Almon, Bert. *Gary Snyder.* Boise, Idaho: Boise State University, 1979.

Baird, Reed M. "Steuding, Bob. Gary Snyder." *Society For the Study of Midwestern Literature Newsletter* 9, no. 1 (1979): 8–12.

Elder, John. "Seeing Through the Fire: Writers in the Nuclear Age." *New England Review and Bread Loaf Quarterly* 5, no. 4 (1983): 647–54.

Faas, Ekbert. *Towards a New American Poetics: Essays and Interviews.* Santa Barbara, CA: Black Sparrow, 1978.

Geneson, Paul. "An Interview with Gary Snyder." *Ohio Review* 18, no. 3 (1977): 84.

Holaday, Woon-ping Chin. "Formlessness and Form in Gary Snyder's 'Mountains and Rivers Without End.'" *Sagetrieb: A Journal Devoted to Poets in the Pound-H,* Spring 1986, 41–52.

Howard, Richard. "Gary Snyder: To Hold Both History and Wilderness in Mind." In *Alone With America: Essays on the Art of Poetry in the United States,* enlarged ed., 562–77. New York: Antheneum, 1980.

Hunt, Anthony. "'Bubbs Creek Haircut': Gary Snyder's 'Great Departure' in Mountains and Rivers Without End." *Western American Literature* 15 (1980): 163–75.

Janik, Del Ivan. "Gary Snyder: The Public Function of Poetry, and Turtle Island." *Notes on Modern American Literature* 3 (1979): item 24.

Kern, Robert. "Recipes, Catalogues, Open Form Poetics: Gary Snyder's Archetypal Voice." *Contemporary Literature* 18 (1977): 173–97.

Leed, Jacob. "Gary Snyder: An Unpublished Preface." *Journal of Modern Literature,* March 1986, 177–80.

McLean, William Scott, ed. *The Real Work.* New York: New Directions, 1980.

McLeod, Dan. "The Chinese Hermit in the American Wilderness." *Tamkang Review* 14, nos. 1–4 (1983–84): 154–71.

McNeil, Katherine. *Gary Snyder, a Bibliography.* New York: Phoenix Bookshop, 1983.

Miles, Jeffrey. "Making it to Cold Mountain: Han-shan in The Dharma Bums." *Essays on the Literature of Mountaineering,* ed. Armand E. Singer, 95–105. Morgantown, WV: West Virginia University Press, 1982.

Murphy, Patrick D. "Mythic and Fantastic: Gary Snyder's Mountains and Rivers Without End." *Extrapolation: A Journal of Science Fiction and Fantasy* 26, no. 4 (1985): 290–99.

Parkinson, Thomas F. "The Poetry of Gary Snyder." In Thomas F. Parkinson, *Poets, Poems, Movements.* Ann Arbor, MI: UMI, 1988.

———. "The Poetry of Gary Snyder." *Sagetrieb: A Journal Devoted to Poets in the Pound-H* 3, no. 1 (1984): 49–61.

Paul, Sherman. "From Lookout to Ashram: The Way of Gary Snyder." *Iowa Review* 1 (1970): 76–80.

Robertson, David. "Gary Snyder Riprapping in Yosemite, 1955." *American Poetry*, Fall 1984, 52–59.

Rothberg, Abraham. "Passage to More Than India: The Poetry of Gary Snyder." *Southwest Review* 61, no. 1 (1976): 26–38.

Steuding, Bob. *Gary Snyder.* Boston: Twayne, 1976.

Wyatt, David. "Jeffers, Snyder, and the Ended World." In *The Fall Into Eden: Landscape and Imagination in California*, 174–205. Cambridge University Press, 1986.